RACECAR

RACECAR

Searching for the Limit in Formula SAE

MATT BROWN

7🚗

DATE	REV.	DESCRIPTION	APPD.
19-OCT-2011	00	INITIAL RELEASE	MB

RACECAR

SEARCHING FOR THE LIMIT IN FORMULA SAE

PHOTOGRAPHS © 2006-2007 DAN GIELAS, BOBBY ALLEY, JAMES WALTMAN, GLORIA BURRIS, AND WESLEY BLACKMAN.

BACK COVER PHOTOGRAPHS (CLOCKWISE FROM THE TOP) BY: DAN GIELAS, GLORIA BURRIS, BOBBY ALLEY, AND WESLEY BLACKMAN.

SEVEN CAR PUBLISHING
WWW.SEVENCARPUBLISHING.COM

FIRST EDITION

FORMAT	ISBN	REVISION	SHEET
PAPERBACK	0-98471-931-8	00	4 of 226

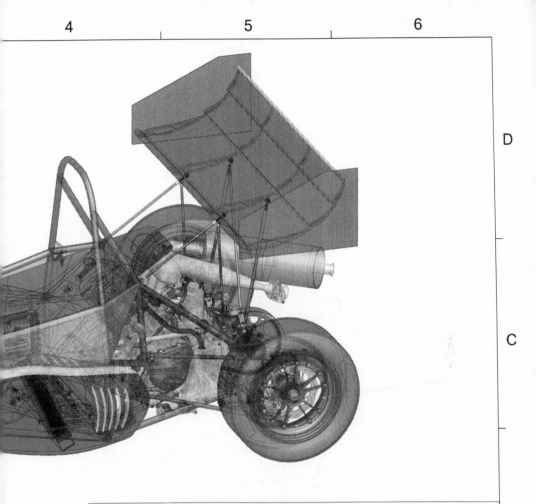

For Mitch and Gloria, Lee and Michelle, and all the other moms and dads who spent their time and money so that we kids could play with our racecars.

Introduction

I'm not the type of person to quit something partway through; I feel that it's important to finish what I've started. I'm also not the type of person to sit down and think about things before I start them. That is, I don't think about how much time and money it will cost me in the end.

Well, that's not entirely true; I usually sit down for about 15 seconds and pull some numbers straight out of my bum. But then I forget to incorporate the rule of Pi. The rule of Pi goes like this: If you think it will take you a week to build a desk, it will actually take you about 3.14 weeks. And if the budget for said desk is $100, you will be out $314 before it's all over.

Auto racing is a little bit different than desk building. The equation for time and money spent on any auto racing venture looks like this:

$$Cost_{Actual} = Cost_0 + \sum_{n=1}^{\infty} \frac{n\pi^4}{5.67x10^{-8}} + nM_{oney}{}^n + \text{ForGoddamnEver} - G_f$$

Where M_{oney} is all the money you will make in the next ten years and G_f is any semblance of a relationship you expect to maintain. (Note: If you substitute G_f with W_{ife}, then you have to add Lawyer$_{Divorce}$ and then M_{oney} approaches an asymptote which is roughly equal to the Gross Domestic Product of Belgium.)

Formula SAE is no different. Most of the money to build the car comes from the school and sponsors, yet somehow, you will have no money. Anyway, the graphical representation looks something like this:

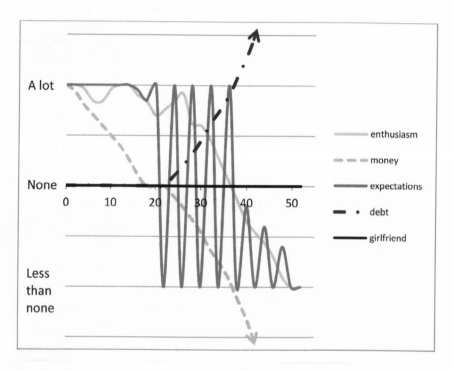

The X axis represents the number of weeks spent on the project. Of course, you expected to be done around week 30, but when 30 comes around and you're not even close, you start to frantically spend money and time trying to finish your project. Expectations swing wildly between "Greatest formula car in history" and "I swear to god I'm going to set this thing on fire and push it into the ocean." Around week 40 your roommate starts

10

to suspect that your wild mood swings are the result of a cocaine addiction, your mounting debt looks suspiciously like you have a cocaine addiction, and you start to wish you had spent all your time and money on something less addicting and less destructive to your health and social life. Cocaine, for instance.

You quit your job eight months ago when last year's car was behind schedule and you "needed" to spend an extra 20 hours a week on it. You could save money on food by eating at home but you don't have time to cook, you don't know how to cook, and your oven smells questionable ever since you used it to bake a carbon fiber steering wheel. Your grades are dropping like panties at all the frat parties you've never been to, and the school wants $1200 for that class you failed last semester because for some reason, financial aid only counts if you don't get an F.

Week 50 rolls around. Your expectations are nil, you can't remember what classes you're supposed to be going to, you start to confuse your net worth with the national debt clock, and you've actually scouted an area in Texas where you can easily roll a flaming formula car off a cliff and into the Gulf of Mexico.

New members ask how they can help. You resist the urge to shake them violently, slap them in the face and yell "Get out while you still can! Don't worry about me, I'm already gone! You have to live on! Live on without auto racing!"

Instead, you hand them a wrench and put them to work. For the last two weeks approaching the race, you spend all of your time on the car. Wake up, work on the car, go to school, come back for lunch and work on the car for an hour, go back to school only to return in five hours and work on the car until 3am when you finally manage to squeeze in a shower before going to sleep. You were supposed to go running today, but you didn't because you were busy with the car. Clean the bathroom? No time, just strap a bomb to a can of Lysol and shut the door.

Everything is behind schedule: the final bits and pieces, the packing, the laundry, leaving. Things on your "must do" list reach a new level of half-assery, and your "should probably do" list might as well include "catch a unicorn" and "fart rainbows" 'cause that crap is never going to happen.

Six hours after your planned departure time, you leave with a hastily packed trailer, thinking about what you might have forgotten and hoping the school doesn't mind too much that you left your half of the shop looking like the Tasmanian Devil had a seizure.

Driving a borrowed van through the desert, you blast "No Sleep till Brooklyn" on the radio and slam three Red Bulls in a struggle to keep your eyelids open. Everyone else is passed out, and you realize that there is a not insignificant chance that the guy in the passenger's seat is actually dead from exhaustion. People are strewn about the cabin like bodies in the wake of an alien invasion. In the back seat, you see a lawn gnome ride an emu over the Georgia Dome before the rumble of the edge of the highway rips you from your slumber. You have no energy left. You've spent the last year of your life giving everything you had. You're out of money, out of sanity, out of empathy, and out of Red Bull.

And the competition hasn't even started yet.

When you spend all of your time doing any one activity, that activity will become more important than anything in the world to you. If you give up so much of yourself for a quest, there will be a part of you which will live in the excitement of, and die in the failure of that endeavor. It is blissful and unfortunate, it is terrifyingly exciting, and it is endlessly adventurous.

Our quest was one of racecars, and it is the same story for anyone who has ever been part of a Formula SAE team. We get addicted. We can't help it. It is part of our love for engineering and our drive to be the best. People wonder why we are obsessed and we wonder why they're not. Someone asks us why we spend 70 hours a week building a racecar and we just look at them like a dog looks at a mime.

Why? What do you mean, Why?
Because racecar.
That's why.

1

Superfast

"I will drive flat out all the time." – Gilles Villeneuve

I slowly pulled forward to fill the space left empty by the dark green Mazda that had just entered the course. The starter, an older heavyset lady wearing a yellow vest and holding a clipboard, was looking down at my front tires as they approached the staging line. When I reached the line, she held up her hand and I jabbed the brake pedal. I took a deep breath and went over the first few parts of the course in my head: a sharp left, then an immediate, increasing radius right, followed by a straightaway into a left turn-around.

The starter diligently watched the car that was on course, patiently waiting for them to finish. The parking lot that this course was laid out in was mostly flat, but our racecar was so low that the Mazda seemed to disappear over the horizon behind a field of orange cones.

Over the low rumble of the engine, I heard a voice over the loudspeaker.

"And coming to the line in the University of Oklahoma car

is... Superfast Matt McCoy."

I laughed quietly inside my helmet, and then brought my attention back to the race course. A sharp left, then an immediate right followed by a straightaway into a left turn-around.

As soon as the Mazda crossed the finish line, the starter pointed at me. I nodded and she waved the green flag. I took a deep breath, revved the engine up, and let out the clutch.

The rear tires slipped just a bit before the car shot forward. I navigated through the awkwardly slow first two corners, and then planted the throttle through the straightaway, carefully listening to the engine for the shift points. Second gear, third gear, fourth gear. The sea of orange cones was a vibrating blur, whizzing by like little orange stars on the deck of the Millennium Falcon.

Our 2006 car would accelerate and corner with more performance than you could get out of a quarter million dollar supercar. It was like being shot out of a catapult onto a rollercoaster. Add to that a screaming engine spinning at 12,000 RPMs, vibrations that rattle your eyeballs, and suspension that transmits every crack in the pavement directly into your spine. Driving a racecar is an exercise in dancing on the edge of control: pushing the tires to their traction limit, feeling the car feedback, reacting instantly, and carefully planning every turn. It's like playing a chess game during a prison riot.

We loved it. We didn't like it, we loved it. The smell of tires and high octane gas, the piercing scream of a high revving engine, the feeling of g-forces so high you think your teeth are going to fall out. For the better part of four years, we dedicated our lives to it. Not only did we spend all of our free time with the car, we spent a good part of our class time and more than a little of our sleep time designing, building, fixing, rebuilding, testing, fixing, racing, and drinking. Er, fixing.

And drinking.

It was early August, three weeks before the start of the fall semester. We had wrapped up our 2006 season just a few weeks before: a year of stress and sleepless nights ending with an 8th place finish. That was a big step up for us; most of our finishes before that were described as "In the top 60." Even with 130 competitors, it just didn't sound particularly impressive.

"Hey, we're in the top 60!"

"Were there 8000 competitors?"

There were not.

Today's race was an autocross; a course laid out with orange cones in a parking lot. This parking lot was huge, built into a gradual hill with separate sections sloping down towards the horse racing track that so kindly let us use their facilities for one Sunday each month. We were only using one section of the lot, at the top of the hill far away from the horse track.

There was no head-to-head racing; autocross is a time trial. In other words, each lap is timed and there is only one car on the race course at a time. There are no employees or stewards; drivers spend half the day driving and the other half running out onto the course to pick up cones that other cars hit.

Most of the cars were daily drivers; Honda Civics, Mazda Miatas, and the like. Some of the cars, like ours, were racecars out getting practice before other, more important races. We were the only Formula SAE team at the event; there were over 300 FSAE teams in the world, but we were the only one left in Oklahoma. Both Oklahoma State and the University of Tulsa had teams in the past, but a lack of faculty support had doomed them into memories.

Before the start of the event, the announcer would ask for drivers to come to the trailer for a work assignment. I would always try to get there first and say "I'm looking for something easy, maybe something in the shade." They would always laugh and, more importantly, give me something easy. I learned that trick from Kyle; he had been racing since he was tall enough to reach go-kart pedals and he knew the ropes.

Today, Kyle was going by his new racing name: Speedy Nuts Kyle. After staying up all night preparing and fixing racecars, it didn't seem prudent to fill out the entry form with the correct information. And so we had Speedy Nuts Kyle, Faster Than You Chris, Bobby Ricky Bobby, and Superfast Matt McCoy.

We had made a few modifications to our 2006 car, including adding wings, and we were there to test the improvements. Also, it was fun. Any excuse to drive. Our wings would, in theory, allow the tires to grip better and let the car go around corners faster, but we weren't going to make that design decision on the next year's car until we had data to prove it.

Engineering is, among other things, the practice of making data driven decisions. A smart engineer can intuitively design something that will work well, but a good engineer will always back up his design with analysis and testing. This is why someone who is not the smartest person in the class can be the better engineer. Smart engineers know enough to design without validation most of the time. It's a good start; an efficient use of time in the initial design phase. But it's not engineering. Good engineers are smart enough to know that they don't know everything, so they make decisions based on data.

We were good engineers, or at least we thought we were. In any case, we were smart enough to know that we didn't know everything, so we tested.

The wings were designed and mostly built the previous year. I had finished up the last bit of manufacturing with some help from other team members and we decided to give them a try. We were skeptical; our average corner speed was around 30 MPH and a trip to your local airport will confirm that wings don't do much at 30 MPH. It was basically decided that we would use them a couple of times for the sake of testing, and then mothball them.

Fortunately we had taken the time to finish them and put them on the car, because our predictions were shockingly wrong.

The bigger the corner, the better the car would stick. I could feel the tires gripping more and more the faster I went. Slowing down was like hitting a wall. When the car was going faster than 60, I could basically stand on the brakes. The seatbelts would push into my chest and my helmet would pull forward like I was being shot backwards out of a cannon.

I blew across the finish line and coasted over to our makeshift pit area.

As I was getting out of the car, I said to Kyle, "These wings are freaking amazing! You can feel the car getting more and more grip the faster you go!"

"Well, you got the fastest time of the day, Superfast," Kyle said with a smile.

"And if the back steps out a little, you can hold it or correct it all day long." I continued oblivious, before stopping and asking, "Did you say FTD?"

He nodded.

I smiled. My first FTD. Of course, I knew as soon as Kyle got in the car he would beat my time, but for a few minutes I had the fastest time of anyone else there. I picked up a bottle of water from our ice chest and sat down to catch my breath.

"Frikin' amazing." I said quietly, looking back at the car.

The wings fit the car well, too. When we put them on the car for the first time it was like seeing a friend with a new haircut that just fits them perfectly; like you could have never pictured it before, but there it is. Bam. Perfect. Now girls might talk to you.

The car was small, only about six feet long. The frame was narrow, and the wheels stuck out about two feet on either side connected by petite triangular suspension arms. The wings filled it out nicely, visually setting it apart from go-karts and placing it firmly within the realm of real racecars. It was always a real racecar, but now it looked like one too.

Kyle asked, "You looked fast, did it feel faster?"

I nodded and put the cap on my water bottle, "You can

really feel it in the braking; as you slow down the downforce goes away and the tires start to lock up. I was sliding around a lot though, so it's hard to say."

Ricky jumped in, "You were almost sideways through that sweeper in the back."

Kyle added, "Yea, we were like 'Is he still on the track?'"

"Yea, that was uhhh… that was a decreasing radius and I was going too fast, so I just carried the slide. What's up with '05?" I asked, shifting my attention.

I noticed Bobby had taken the intake off and was peering into it with the help of his pocket flashlight.

"Huh?" Kyle asked before answering, "Oh, it won't start."

"Didn't Bobby do a run already?" I asked.

Ricky replied, "They got it started once. It will stay running once you start it, but they can't get it to start again. I told him it's an intake leak, we had this problem before."

Ricky was great at finding the causes of problems and fixing things. He was smart, and though he wasn't arrogant he could sometimes be stubborn in the face of opposing ideas. He also had a lot of good ideas, which is why we called him "Good Idea Ricky." We didn't actually call him that, but after every good idea he would announce "That's why they call me Good Idea Ricky." He wasn't driving today, so his motivation to come up with good ideas to fix the '05 car was limited to a suggestion.

Chris was gone, likely being pissed off somewhere for spending $30 on an entry fee and wasting a Saturday to come out to the race and not get to drive. I understood his frustration, we spent too much time on these cars not to drive them every chance we had, and for those of us who had quit our jobs to spend more time on the car, $30 was a small fortune. Ramen noodles like whoa.

Kyle and I were driving the 2006 car; Bobby and Chris were supposed to be driving the 2005 car. Our 2004 car had long since fallen into disrepair and disassembly. When crunch time comes and the new car needs to be finished, it's all too easy to pick parts off the old cars. Eventually the old cars become old car frames, too broken to be fixed. '04 was never very fast anyway, though it was loads of fun.

The first time I drove one of our racecars was the summer of 2004. It was the first time I drove any kind of racecar at all. I wasn't sure how I should drive it; I knew that the pedal on the right made it go, and the pedal on the left made it stop, but I wasn't sure about much more than that. Don't hit the cones, right? When do I shift? Do I need to talk to the TV camera about the cool refreshment of Arctic Rush Gatorade?

So, I just put the pedal on the right to the floor and drove it like it was stolen. I threw it into the corners too fast and the back tires broke loose. I counter-steered and carried the slide too far, all the way through the next corner in some places. After three seemingly insanely fast laps, I brought it in and asked Kyle, "Was that too fast?"

He just laughed and said, "No."

One of the first things you learn when you start racing cars is that the fastest way around a track doesn't feel like the fastest way around that track. If it feels like you're going fast, you're usually just sliding sideways and scrubbing off all your speed. This is a lesson I didn't learn until well after that day in 2004. That day I just drove like a mad man. I knew I could be a fast driver; I knew I could be competitive. What I didn't know was that I was driving like an idiot. It sure was fun though.

That was when the obsession started for me. Obsession bordering on addiction: one time is all it takes and you're hooked. Addicted not only to driving, but to staying up all night making the car faster. Faster, and then faster still, resisting the almost overwhelming urge to pawn your mother's television so you can score some titanium intake valves.

I drove the car every chance I got after that. Whenever the

21

team had a scheduled driving day, I would try to get someone to take my delivery shift at the pizza place. Sometimes I would get off early and try to get in a lap or two before the team packed up for the day. One time I stopped by and drove a couple of laps while I was on a delivery.

I thought it was better than sex, but that might have been because I was spending all my time on the racecar and never actually having sex. I tried to learn how to drive every chance I got; I would read books, look for web pages, and ask other drivers for advice. There was surprisingly little information out there. Kyle would give me some generic advice whenever I asked him, but I could never explain what the car was doing, or how it felt.

"Stay closer to the slalom cones, everyone takes them too wide. And be early, you should be turning in just as you get to the cone, not after."

That was the best piece of advice I got from Kyle on driving. It was true, everyone drove way too wide through the slalom, like they were afraid of the cones. I would always focus on the slaloms, trying to find that edge of as-close-as-possible without knocking the cones over. A perfect slalom would be driving over the base of the cones, making them wobble a bit, but not fall over. Ricky called it "The butter zone."

By the beginning of the 2007 season, I was faster than everyone on the team except Kyle. I could edge out Bobby, Ricky, and Chris by a few tenths of a second, but never Kyle. When I first started racing, he already had several years of experience in several different racing series. For the previous three years, I was getting better and better, but so was he. When I first started driving the car, I would look at the lap times and say, "Man, Kyle is three seconds ahead of me." Eventually it became, "Man, Kyle is three tenths of a second ahead of me." I could never figure out where he was making up all that time, in either case. I thought that I would never beat him. There was a local autocross race one weekend in late 2005 that Kyle couldn't make it to. When

Chris asked if anyone wanted to go, I jumped at the opportunity; I figured it would be the only chance I would ever get to come in first place. I ended up second place behind Chris, and went home disappointed and trophyless.

"You're up next, let's go!" The starter lady yelled in my direction.

I jumped back in the car and started getting ready. It was a lengthy procedure, more like preparing to fly an airplane than driving. I sat forward and reached behind me to get the two lap belts and then over my shoulders for the shoulder belts. Then I sat back and reached between my legs for the "submarine strap" and attached them all. I pulled the helmet on, strapped it, put the steering wheel back on, tightened the seatbelt straps, pressed the arm button, pulled in the clutch, started the car, and shifted into first gear. By the time I was done, the starter was impatiently waving me to the staging line and as soon as I got there she gave me the green flag.

During my second lap, I drove a little more carefully and went even faster. I got the perfect exit to the last corner, planted the pedal to the floor and watched the finish line accelerate towards me. Second gear, third gear, fourth gear. "AhhhHHHH Ha HAAA!" I yelled in my helmet. Then, only a few feet from the finish line, I saw an explosion of black plastic come from the front of the car. The front wing had collapsed under the aerodynamic force and shattered into half a dozen pieces. The lap time was again the fastest time so far, but the front wings were done for the day.

As soon as the wing broke, all the corner workers waved their red flags to indicate that the event was stopped. I pulled up next to Ricky and Kyle and got out of the car. We all stood looking with disappointment at the car as one of the course workers ran over with the pieces of wing he had collected off of the course.

"That was pretty spectacular," the course worker said with a smile. "It just blew up right there at the finish."

"Man, I really wanted to drive with the wings," Kyle said quietly as he started to take the rear wing off.

Bobby had given up on '05 and helped Kyle remove the rear wings while I pulled off the front wings. I got back in the car and waited for my next run, frustrated, excited, and exhausted. I still had one run left, and Kyle had his three after me.

I finished my last lap more than two seconds slower without the wings. I got out and sat down in the shade, upset that the car was broken and that I didn't get another chance to improve my lap time.

"Coming up to the line in the University of Oklahoma race-car is... Speedy Nuts Kyle."

A laugh pierced through my grimace.

Kyle got the green flag and took off. He slid around a little at first, feeling out the course, but by the third corner the car was winding through the cones like a slot car. He was the tallest person on the team, but he was also skinny so it didn't really affect his lap times. In a racecar you always want to be as light as possible and have all your weight as low as possible. The driver is one of the easiest ways to change that, but when everyone is an amateur, driver skill is often a much bigger factor than weight.

When he came across the finish line, the timing marquee lit up to show his time. He was almost two seconds slower than my fastest lap.

"Did it feel like it was pushing in the slow corners?" he asked me, still in the car.

"No, I... I don't know." I said.

Kyle noticed minor changes in balance, even through a single corner. For me to notice a balance change, there pretty much had to be a fat guy sitting on the nose. I could react to changes in the balance of the car, but I couldn't explain them.

Every time I drove before him, he would ask "How does the car feel?" seemingly expecting me to say something like "Well, coming into corner four it breaks out a little with a push at the apex once the decreasing radius starts." My responses, however, were more like "It's freakin' fast, dude." But for some reason, he always asked me.

He went out for his second lap and finished a second and a half away from my time.

His third lap was a little faster, but not enough to beat my fastest time. For the first time ever, I had beaten Kyle. Granted, it was only because the wings pretty much disintegrated before he could use them, and yes, his no-wing lap was better than my no-wing lap, but hey, who's counting? For the rest of the day, no one else could beat my time in any car. Superfast Matt McCoy was undefeated. Also, and probably more importantly, we knew that the wings were effective. The Formula SAE competition would be a much bigger challenge than beating up on Mazda Miatas in a parking lot in Oklahoma City, and we needed every advantage we could get if we were going to win.

We hung around for the trophy handouts so I could pick up my FTD trophy. The "trophies" were just refrigerator magnets that said "1st place" or "FTD", but they were neat to collect and they only gave one FTD at each race.

"You guys want to get some dinner?" Kyle asked.

"I need to get home, I told my girlfriend I'd be home an hour ago," Chris said.

"Yeah, I gotta get home and do some homework," Bobby replied.

"You're not going to do homework, you never do homework," I said.

"Yea, that's the problem," Bobby said. "I have to start or I'm going to be in college forever."

"Oh, that's not true," Chris said in a sarcastic voice. "Eventually they'll kick you out."

"Well I could go for some dinner," I said.

I was riding with Kyle anyway, so it seemed appropriate. It

was too cold that morning for a motorcycle, and my truck was not the kind of vehicle to trust for trips longer than five miles.

"I wish I would have gotten a chance to drive with the wings, it looked really fast," Kyle said quietly as we walked to his car. He laughed and added, "It looked like an explosion, pieces just went everywhere."

"Eh, we picked up all the bits. We'll have it back together in a day or two. I'll add some fiberglass reinforcement so it doesn't collapse again."

As we drove out of the parking lot, Kyle said, "Congratulations on your FTD."

"You know, I never thought I'd get an FTD," I said, looking down at my shiny new refrigerator magnet. "Maybe next time I can do it without destroying the car," I added with a smile.

2

The First Meeting of the Year

"A problem well stated is a problem half-solved." – Charles Kettering

There were 2416 parts on our 2006 car. That doesn't count any of the engine internals or the individual components in the dampers, or any other system subcomponents that we didn't make. That's just the parts we chose and designed into the system. Some of them were made, some purchased, some modified, but all of them were on the car for a specific reason.

The bolt that held the front upper control arm to the frame was a 1/4" diameter, 28 threads per inch, 1-5/32" nominal length AN bolt with a deformed thread lock nut. We could have changed any of those variables and changed the efficiency or effectiveness of the bolted joint. Not only the joint, but also the front suspension subsystem and the entire vehicle system. We could probably have used a smaller, lighter 10-32 bolt, or a titanium bolt of the same size. We could have used cheaper bolts that are not as strong, we could have pressed in a pin and not used a bolt at all, or we could have just jammed a screwdriver in the hole and called it a day.

Every component on the car has tradeoffs, and nearly every part of the car affects some other part of the car, so the possible combinations of functional designs are basically infinite. Common sense will knock out most of the possibilities. The rules narrow the field a bit, and experience with racecars will give some direction. Basic engineering practices dictate a more refined design direction and eventually you get to a manageable number of designs for each component. Once you've selected probable design directions for the component, you have to see how those designs affect other systems, whether or not those designs are in the budget, and how serviceable and manufacturable the designs are. And of course every once in a while, someone says, "Hey! Why don't we stake in sphericals, machine aluminum pickups, and use shims to adjust camber!" and you have to start the whole process over again. Finally, once you've narrowed it down to what you think is the best design for that component, all you have to do is repeat the process 2415 more times.

To be fair, in a lot of cases you can just copy what the team did the previous year or what some of the other teams do, but the challenge of making it lighter, stronger, better, and faster is why we joined the team. The whole point of racing is to win, so good enough is never good enough. We had to make it better, or else we would never be faster, we would never win, and we would spend the rest of our lives designing toasters, wondering what we could have done, if only we had tried a little bit harder.

Also, I should point out that if the aforementioned front upper control arm bolt were to fail, that corner of the car would become useless and you would have a three wheeled out of control machine being piloted by an amateur driver in a school parking lot at 75 miles per hour. And, of course, all of this happens while the vast majority of students are completely oblivious to the fact that there is a racecar on campus and that it could, at any moment, come crashing through their classroom wall sans a right front upper control arm.

So selecting, manufacturing, and assembling 2416 components all connected together in some semblance of a racecar is difficult at best and requires more time and effort than any reasonable person would expect from a group of college students. Knowing that, I was astounded by what some of the other teams had done when I finally made it to my first competition in 2005. You spend a year of your life working on these cars, pouring in all your time and energy, only to show up at the competition and realize that you are competing against a thousand other people that had done the exact same thing. Some of them with more team members, more resources, and more support from their faculty and school. There were teams that had done detailed design studies on things I had never heard of. Those were top schools with serious resources and dozens of very hard working team members.

Cornell had been the top team for several years. They were an Ivy League school with a faculty advisor that was very involved with the team. Faculty involvement was one of the most important things for a team to be competitive. Having a dedicated faculty advisor gave the team continuity and direction, and kept them from making the same stupid mistakes every four years. Penn State was another top team; with clever designs and an abundance of magnesium and titanium, they managed to get their car weight down to 350 pounds, nearly a hundred pounds less than the weight of the motorcycle that they got their engine out of. These are small cars, yes, but they do have four wheels, an engine, a fuel tank and all the other components necessary for a racecar. Western Australia had students on their team that were doing their PhD theses on parts of the racecar.

The University of Wisconsin-Madison always had a good showing. They had a relatively simple design, but they knew what they were doing and they did it well. At the 2006 competition, I had worked for several weeks developing a report and presentation for the engine design award. I got second place behind

Wisconsin. Since then, they had been our nemesis. Every time we heard their name, someone on our team would mumble "Fucking Wisconsin." Usually me. They didn't know that they were our sworn enemies, and they probably never thought twice about Oklahoma, but from our side it was a bitter rivalry.

All of these teams, and a few others, were beyond impressive. They were past college level; they were professional racing teams in a college competition. Our car and what we managed to do on our team was remarkable, but I remember my first competition looking at the cars in the design finals event and thinking that they were light years ahead of us. They were world class universities and we were just the University of Somewhere-In-The-Middle-I-Think-Near-Kansas-Maybe.

For most people, it took several weeks for the scope of the project to sink in. That was probably a good thing, because if some people had known what they were getting themselves into, they might have never shown up.

In any case, new members always arrived with bright eyes, ready to tweak this baby like it was their bitchin' Camaro. And it was our job to calmly explain that we would carefully design, build, test, refine, and race a purpose built racing car from the ground up every year. We usually left out the part about it taking over your life. They would figure that out soon enough.

"I think this is one of the most exciting things that you can do as a student."

It was the first official meeting of the year and, as the Chief Engineer, I was giving my brief intro speech. Our meetings were held in one of the newer conference rooms in the engineering building. We were told that we were allowed to use the room as long as we were careful with the chairs. The chairs were very important, for some reason.

The Public Relations team, or Ricky as he was more commonly called, had put up posters around campus and given presentations in engineering classes in an effort to get more people to come to our meetings. It seemed to have worked; the room was more than full. Ricky had been our token Electrical Engineering major, but for the '07 car he had decided to trade in the wires and electrical tape for fliers and sponsorship presentations. Most of the core team members, including me, were graduating at the end of the year and so we were trying to recruit as many new members as possible to ensure the team would be successful in the future.

"There are about… fifty people here. Next week there will be thirty, and then ten, then about four of you will stick around for the rest of the year." I stated matter-of-factly, thinking about how the rest would spend their free time playing Mario Kart, taking bong hits, and complaining about how hard college is.

"If you want to get into automotive or auto racing, you should be one of those four people. It's hard work, but it's the most fun I've ever had. Most of the design work on the '07 car is finished, but there is still a lot of work that needs to be done. I have printed a list of new member projects that you need to complete to become a member of the team. They are all basic stuff, but you need to be able to show us that you can complete a project. This will also get you in the machine shop and the computer lab and get you involved with the other team members."

I took it upon myself to try to make it as easy as possible for new members to get involved and I expected that it would be easier if they each had a project to show off their skills and work ethic. I had a difficult time becoming a team member when I first joined. I had showed up to meetings for over a year before I got handed any projects. I was, admittedly, quiet and shy about the whole thing, but so were most of the good team members. I wanted to be a part of the team, I wanted to be the Chief Engineer since the first day I saw the car on the lawn outside the engineering building. I showed up week after week for over a year, grinding and cutting and sweeping the floor. I didn't get to design anything, so I spent the first year and a half watching the older team members make terrible design choices. James Joyce once said that mistakes are the portals of discovery, and that can be as true for spectators as it is for the person making the mistake.

I tried to learn and make the best of it but it was irritating to watch the team leaders pick team members to go to the competitions who were less knowledgeable and less hardworking than I. I have to admit that I held a resentment towards many of the old team leaders and avoided them when they came back to visit after their graduation. My attitude was probably immature and unhealthy but, in any case, I was going to make sure that didn't happen to any new members this year.

I turned to Bobby and added, "That's all I got."

Bobby was the Team Captain and was a relatively young team member for that position. He had moved to Oklahoma from Bakersfield, California and was living in his grandparents' guesthouse. Bobby was short with dark hair, a few of them turning grey already. He had been in Army ROTC but stopped going so he could spend time on the race team. Most people on the team had bad grades and failed a few classes, but Bobby excelled at it. I wasn't sure if he would ever graduate. We joked that he would be on the team until the sun went supernova.

"Okay, so we'll go over some basics about us so you all know what we do here. We are the Sooner Racing Team. I am the president, Bobby. Matt, who we just heard from, is the Chief Engineer. Andy is in charge of the chassis," Bobby held his hand out, gesturing towards Andy and Chris, "Chris is in charge of the engine. We have two faculty advisors, Dr. Sommer and Mitch Burris, both of them are in the back there. We compete in Formula SAE. FSAE is an international intercollegiate competition. It is hosted by the Society of Automotive Engineers, that's where the SAE comes in. We build a small, formula-style racecar powered by a 600cc motorcycle engine with a twenty millimeter inlet restrictor. The rules are pretty open for racing, so we can do a lot of things. There are about three hundred teams around the world. We compete in Detroit against a hundred and thirty teams and in California against eighty. Teams come from all over the world, England, Australia, Finland, Japan…"

I could tell people were already losing interest. I looked around and made mental notes of the people that were still paying close attention. Our meetings were not Michael Bay movies, but they were about racecars, and if that's not enough to keep someone awake then they were probably at the wrong meeting.

Bobby continued, "The competitions have several different events including an autocross race, an endurance race, acceleration, and a skidpad. There are also static events like design where professional racecar engineers ask a lot of hard questions and rank the teams. The team members are the ones that drive the car, and if you put the time in, you'll get your chance in the car," He paused, trying to remember if he hit all the points, "Like Matt said, it's a lot of fun and, uh… Oh yea, you all need to fill out the new member form and the sign in sheet. Who has the sign in sheet?" We all looked around. One of the new members picked it up and waved it in the air. "Okay, he's got it. Make sure you sign it before you leave. So now Chris and Andy are going to give a, uh, introduction to their systems."

When designing something like a racecar, you have to take a systems approach. That is, you have to look at the whole system. You can't just build a component, bolt it on, and expect it to work, no matter how well designed it is. All the cooks in the kitchen should be talking to each other, otherwise you end up eating mango lasagna with carrots and cake icing. It is the well-engineered and thoughtful interaction of all the components that makes a good car. This is why you can have 15 completely different vehicle designs that all perform at the same level. Every component affects every other component, and they all need to work together. This is as true about the car as it is about the team that builds the car.

The two system leaders were in charge of coordinating components within the two major systems: chassis and powertrain. The Chief Engineer was in charge of, among other things, coordinating the two major systems. This meant lots of communication with Chris and Andy.

Andy was easy to communicate with. He was straightforward and organized, and while he sometimes got defensive, he was never offensive. He was pale and soft spoken, and he had this "nice young man" demeanor about him which was mostly accurate except that sometimes he would go on angry profane rants about seemingly random stuff. They were never directed towards any one of us, and they were always hilariously entertaining.

Chris was less consistent. He was fun to be around in social settings; if he was talking, drinking, or generally not busy, he was usually goofy-happy. During team meetings and while manufacturing, he was serious and direct, bordering on rude. Sometimes he could be stubborn about design ideas. His ideas were usually as good as any, so it rarely became an issue. In any case, we couldn't do it without him, and I told him that at the end of the previous year when he was thinking about leaving the team. He worked very hard, spent a lot of time, and was very good at manufacturing. There were always about five people every year that spent way too much time with the team, and he was one of them.

Both Chris and Bobby were short and somewhat stocky with dark hair. You could be forgiven for mistaking them as brothers. They both had varying lengths of short facial hair, but so did most of the team. Shaving sucks *and* it takes time. Two strikes and you're out.

"Who wants to go first?" Bobby asked.

Chris stepped forward, "I'll go first. So ahh... I'm doing the powertrain this year. The powertrain consists of the engine, which we don't modify, and the drivetrain. We build the intake and exhaust ourselves, and we wire up a fuel injection computer and tune it on the dyno. Let's see... the rear end, the axles, basically everything between the engine output and the rear wheels is part of the drivetrain. Like Bobby said, we use a 600cc motorcycle engine..."

After Chris had finished, everyone was silent for a moment before a hand went up.

"You," Chris pointed towards the hand.

"Why don't you do anything to the engine, like port the heads or change the cams?"

Chris squinted and looked at the back wall. This was his I'm-explaining-something pose.

"With the restrictor we flow less air than the stock bike, so porting the heads would actually hurt us," Chris responded.

Good answer, I thought.

Chris asked, "Anyone else?"

No response.

"Okay, Andy," Bobby said.

"Well, I'm Andy, and I'm doing the chassis system this year. The chassis—"

"Louder!" Kyle shouted from the back.

"Huh?"

"Talk louder, we can't hear you."

"Oh, okay," Andy said quietly, and then raised his voice, "I'm Andy and I'm doing the Chassis system this year. The chassis is basically everything... well it *is* everything that isn't powertrain. So the suspension, and the frame, and the body, and the aerodynamics. Well, Matt's actually doing the aerodynamics," Andy's voice had slowly faded back to its original volume, "So, that's about it. Any questions?"

Andy was not a master of holding an audience captive. Then again, this wasn't a communications club.

Bobby stepped back up to the podium, "Phil is in charge of the shop, making sure everything is stocked and all the equipment is maintained," Bobby looked at Phil and quietly asked, "Do you want to say anything?"

Phil was a large guy, and he always wore the same things: a

plain white t-shirt, black hat, blue jeans, and prescription sun-glasses. He was like a cartoon character, and I imagine his closet was just full of white t-shirts and blue jeans all comically lined up.

Without moving from where he was standing, Phil gave a short introduction, "The only thing that I ask is that you clean up after yourselves and put your tools away. I am not the jani-tor; it is not my job to clean up after you. And try not to break anything, if you break our equipment, or our jig table... well if you break the jig table I'll be impressed, but then I'll ask you to leave." Phil punctuated his statement with a dorky laugh. He always had this laugh that he did, like he was trying to disarm people after everything he said. It was kind of goofy, but I ap-preciate happy people and Phil was almost always both of these things and the laugh just showed it.

Most of the system designers had been on the team for a while. Kyle was designing the brakes and had been on the team for four years already. Chris had also been around for four years and Bobby had been around for three. There were a couple people that we had recruited at the end of the previous year for design positions because we were short on people and needed designers to begin designing their systems before the school year started. Ralph was a graduate student who had expressed some interest in the team, and we had given him the task of designing some of the suspension components. Nobody knew Ralph, but he was in-terested in the position and we needed people. We were handing him a relatively important piece of the puzzle, so we did actually explain to him that it would take a lot of time.

Josh had shown up to a few meetings and helped with some things the previous year. In some ways, Josh reminded me of my-self when I first started the team. He had a cheap car that he was always tinkering with, like adding carbon fiber pieces and boring out the throttle body. I had done the same sort of stuff with my cars, not because I expected them to be fast, just because I liked car projects. He showed up to all the meetings, and he was a little socially awkward, partly because he was zoned out a lot. I did the same thing, constantly dreaming of designs for the car. I knew it was hard to be a part of the team when you're not great at socializing, so I tried to make sure Josh had opportunities to work on the car. It also helped that he was good with electronics because we needed an electronics guy. So we put him in charge of the electronics system.

"Does anyone have anything else?" Bobby asked.

"One more thing," I stepped up to the podium, "This is for the designers out there, not so much the new members, but... I guess you can be a part of it too if you want. I know the Chief Engineer has the distinction of being the asshole." I paused for comments, but got none. "So in an effort to have more positive reinforcement, which I'm told is a better motivator, I've decided to introduce Mr. McCoy's Wall of Stars and Cars." I held up my sheet of paper with everyone's name on it, and a little gold star sticker by each name. "This is sort of like kindergarten. When you do a good job, like finishing a task on time, you get a gold star. If you do something really great, like finishing ahead of schedule and beyond your performance goals, you get a car sticker, which is worth five stars. Whoever has the most stars at the end of the year gets a cookie. One of those big pizza sized cookies you get at the mall." Everyone had a look of amusement. I finished my brief presentation with a thumbs up and the cheesiest smile I could muster. "And since you all showed up and have taken some responsibility, you all start out with one star."

My roommate, a business major, had told me about a class he was taking and the techniques they taught for motivating employees. Having a friendly competition where the reward is not inherently valuable but where the participants feel connected to the competition, was one technique that had a lot of positive results. We were all into racing, and thus inherently competitive anyway, so that part would take care of itself. The challenge was presenting this competition to a bunch of cynical college students without it coming off as a cheesy motivational technique from a business class. I decided that the best way to accomplish that would be to present it in the most childish way, coming off as an obvious cheesy motivational technique from a business class.

This was my management technique. When I first got involved with the team, I had decided to pursue Chief Engineer, and had made that well known almost a year before the elections. I chose Chief Engineer instead of Team Captain because I wanted to be more focused on engineering than management. However, it had become obvious over the last few months that Chief Engineer was more Chief than Engineer and I would be managing whether I wanted to or not, and whether I was good at it or not. So I did some research, talked to some people, and this was my idea. I picked up some other good stuff: don't criticize teammates in front of other team members, don't criticize at all without adding some praise in as well. A bunch of stuff to protect a person's ego, and a bunch of stuff that's hard to do when you're supposed to be the asshole.

Bobby stepped up to the podium, "Okay, well, we will be driving the '06 car on Saturday at about noon, everyone is welcome to come out and watch and help us set out cones."

I leaned in and quickly added, "Gold stars for everyone who shows up to help!"

Bobby continued, "We'll meet at the north shop at noon. There's a map on our web page. Aaaand that's it for our meeting."

As people were shuffling out, I picked up the new member sign in sheet and scanned the "System you are interested in (chassis / powertrain)" column. Some jackass had written in "driver."

Every year, among the crowd of would-be team members, we had the same personalities show up. They were always different faces, but they might as well have been the same person. For some reason, there was always a jackass; we would basically ignore him and he would be gone within two weeks. On the other side of the spectrum, we always had a couple of people that were way too interested in the team at their first meeting and told us all the stuff they wanted to do. We liked enthusiasm, but these people always lost interest in the team by the third week. I always tried to get these members involved, hoping they would stay, but they never did. It was just in their personalities. It was probably just as good, those people usually wanted to do esoteric or over-complicated designs and we always discouraged this "outside the box" thinking. It is imprudent and unwise to think outside the box until you've explored all the possibilities inside the box. We were still inside the box. It's a big box.

There were a few people who would show up to the meetings, help us with whatever needed to be done, and learn about the competition and the cars by reading books and asking intelligent questions. Those people made it onto the team, and I was patiently waiting for the third meeting to see who I would be spending all my time with over the course of the next year. By the time we realized who those people were, I had already learned their names and forgotten them three times. I didn't put too much effort into remembering names the first few weeks because I knew that I would never see most of those people again.

The team was a close group, but we tried to be welcoming to just about anyone. Anyone who wasn't a jackass.

There was a university policy stating that students could not spend more than 10 hours per week with any university club. I don't think there was a single week where I spent less than 30 hours doing something with the car or the team. There were several hours spent each week on things that never even made it on the look-what-I-did list, stuff like organizing tools and sweeping the floor. I expect it was the same for most of the other dedicated team members. I don't think any one of us really understood how much everyone else did, nor did we wholly grasp the amazing complexity of the cars. Not only were there thousands of components, but many of them were adjustable making for countless setup combinations.

The person behind most of those details on our team was a guy we called Superdave. We called him Superdave because he put us all to shame with the hours he spent on the car. He meticulously picked and designed hundreds of components, and spent hundreds of hours creating a computer model so detailed that it had oil in the engine and water in the radiator. He would show up randomly during the night or day and work for hours on end. I don't think he ever slept. Dave wasn't a very good team player, rarely showing up to the meetings and always refusing any official responsibility, but he did more for the car than most. He didn't usually bother to show up for class either, which is why he was in his seventh year of college and still working on his Bachelor's degree. Somewhere between working on the car and not sleeping, he found the time to go to the gym; he was in better shape than anyone else on the team. He rarely talked, usually walked with his head down and headphones on, with a cup of coffee in his hand. He always had coffee. When he did talk, it was quietly and always about something technical. The guy was a machine, a racecar building robot with no emotions and the innate ability to redesign anything on the car to make it better.

"Did you guys see Dave at the meeting?" Bobby asked.

"He left after the first part of the meeting," I said, "He's probably in the computer lab."

Bobby shrugged his shoulders, "As long as he's working on the car."

I went to the computer lab to see what Dave was up to, but he had already left.

The lab was empty and the fast computer was sitting idle in the corner, beckoning me. I told myself that I wasn't going to work on the car after the meeting, but I had ideas bouncing around in my head which I wanted to get on to the computer. Besides, I knew if I went home I would get ten minutes into homework before getting sidetracked with racecar ideas. Differential Equations would just make me think of limited slip differentials, then the paper would be half calculus and half sketches of stub shaft cross sections, and then I'd just have to rewrite all the homework.

No, staying in the lab would be a better use of my time. I could get the ideas down on the computer and get home by eleven.

Our clutch lever was designed by me two years prior and needed some improvement. The design itself was genius, if I do say so myself. It was steering wheel mounted and allowed the driver to pull the clutch with either hand at any steering angle. It stayed out of the way until it was needed, and then it was right where you expected it to be. The levers were machined aluminum, and while they were easy for us to manufacture, they didn't have the stiffness we wanted. I didn't want to just use a thicker section and take the weight penalty, so I was looking into different materials. Magnesium would have been nice, but we didn't have the machining capabilities or experience for mag. Still, I did analyses on a few different designs; if it was a major advantage, we would have paid someone else to make it.

I watched the clock creep past 11pm, but I still had ideas and I didn't want to lose them. Mag wasn't looking great, but I hadn't yet looked at a hollow steel section. Steel has the same stiffness-to-weight as aluminum, and while it is harder to machine, we could find some off-the-shelf tubing near the size we needed and weld up a decent design pretty quickly. Plus, with a hollow, deeper section we would get much higher stiffness. After some iterative analyses, it looked like a promising design. I loaded the design into the 3D car model and ran it through a kinematic analysis to make sure it was still clear of other parts and still in a user friendly place. We would have to build a mock up to make sure it was comfortable and out of the way, but my design was simple enough that it would probably be just as easy to build it and test the actual handles.

It was late, but the design was still getting better. I continued to iterate, running through the loop of 3D design, stress and stiff-ness analysis, and kinematic analysis until I had something that I thought was great. At 3am, Differential Equations homework was no closer to being finished, but the racecar was a little bit better for my time spent. It was worth it, but my focus was wan-ing and 11pm had long since passed. I saved some screenshots to show to the team, and shut down the computer.

The cold early morning fog made my pending motorcycle ride home seem somewhat unpleasant, but probably for the better since the chill would keep me awake. I thought again about how we should have a bed in the computer lab. Dr. Sommer would never approve, but it would make us so much more productive; I could work for another 45 minutes and also get enough sleep before class. I still had ideas. Ideas that would make the car even better. But sometimes you have to know when to leave, and some-times that *when* was four hours ago.

3

A Change of Plans

"The crashes people remember, but drivers remember the near misses." – Mario Andretti

My eyes darted from the speedometer to the highway in front of me to make sure there was no traffic ahead, and then back to the speedometer, waiting for the needle to move past 112. I was latched on to the motorcycle tightly, like we were shrink-wrapped together. My elbows were pulled in close, my head tucked down, and my eyes peering over the gauge cluster. For most new motorcycles, especially the sport bikes, 112 was a small step on the way to some much more exhilarating number. But for all my effort, that was the limit for my thirty year old, flat black Yamaha. I had milled the block to get better compression, ported the heads, swapped out the mufflers and air box for more flow and removed the turn signals and one of the mirrors for better aerodynamics, but all I could get was 112.

I liked to see if I could go any faster in different weather. If it was colder, the air would be more dense which would cause more aero drag and slow the bike, but it would also increase the horsepower because of the denser intake charge. I wanted to see which one would win out; it was sort of like a guerilla engineering experiment. But alas, 112 every time.

I peeked over at my rearview mirror to make sure I hadn't aroused the interest of any law enforcement officials, then sat up and slowed down to a normal highway speed before taking the exit to the shop.

Our shop was about ten miles from campus in a small commercial building park surrounded by empty fields. To give you an idea of the area, one of our closest neighbors was Toby Keith and his 160 acre ranch. We got away with pretty much everything after business hours because it was so isolated. The shop was part of a large white metal building that we shared with a plumbing business and a masonry company. It was a basic prefab metal building with a small office where we kept the couch (mostly for sleeping) and a bathroom that looked like it was used for cage fighting matches between a homeless man and two-thirds of a broken motorcycle.

Andy and Dave had been there most of the day, Kyle and Ricky had arrived just before me, and Bobby and Chris were there as well. That made up the core team, or most of it. A quorum at least, enough of us to make a big decision. Good thing, too because Andy had bad news.

"We can't package the 600RR," He said, sitting on the welding table with a look of preemptive defiance. "We just can't do it without throwing everything else off."

The 600RR was the newer version of the engine we had been using for the past five years. It had a few advantages: it was a bit lighter, had bigger exhaust valves, and most importantly, we had two of them we got free from Honda.

Unfortunately, while being a great design for a motorcycle, it didn't lend itself well to a racecar. The transmission was stacked vertically, which not only made it top heavy, but also created loads of difficulty in designing a frame and suspension around it. Andy and Dave had been trying to design around it for the past

44

several weeks, apparently to no avail.

Andy continued, "We can't get the suspension geometry we want with it, and to get the frame around it we're going to have to have tubes way out on the side."

"Don't we have a design freeze scheduled in like, two weeks?" I said.

A design freeze is when you stop making changes to the design of the car and you start building it. It is important for a few reasons but the biggest for us was to force everyone to move on to that next step. If you never gave an engineer a deadline, they would just iterate ad infinitum, continually improving the design ever so slightly until the end of time. That's where project management comes in.

"A design freeze doesn't do us any good if the design is bad," Kyle said.

I responded defensively, "I understand that, but we can't just give up the schedule."

"Are you asking if we can go back to the F4i?" Bobby asked Andy.

"I think we have to, unless we want to have a less competitive car," He answered.

Bobby started to make a comparison, "So the advantages of the 600RR are, we have two of them, and... they're a little bit lighter?"

"And slightly larger exhaust valves, but we're restricting the inlet anyway, so that probably doesn't matter," I added.

"And with the F4i we know it very well," Bobby continued.

Chris jumped in, "And we can still use all that dyno info Matt got last year."

"Eh, honestly it would probably all be the same for the 600RR; those two engines are so similar," I said, thinking out loud. "I think performance and tuning wise they're about the same. It really should come down to suspension and chassis. What kind of performance loss are we talking about in those areas?"

"Dave!" Andy shouted across the shop. "How much worse is the 600RR frame?"

Dave walked over and quietly said, "Our torsional rigidity is reduced by about twenty percent and the weight is going to go up by about ten pounds."

That was one of the best things about Superdave; he could quantify everything.

"Ten pounds!?" Kyle leaned forward, jaw open.

Ten pounds was it; that was the end of the 600RR. Weight was vastly, hugely, monstrously important to us. You would have an easier time convincing Sarah Jessica Parker to put on ten pounds. We had spent hundreds of dollars in some areas to decrease weight by a fraction of a pound.

I asked Andy, "And you're okay with redesigning a new frame?" He started to respond but I interrupted with an important qualifier, "In a relatively short amount of time?"

He nodded and said, "Yea."

The car was supposed to have been mostly designed. I wasn't happy, but I also knew that it was the right thing to do. It wasn't anybody's fault; Andy had been working his ass off on the frame and had come to the inevitable conclusion. Designing a frame and suspension is not an easy task by any means. Every little change will affect twenty other things.

Andy had spent hundreds of hours designing the car around the 600RR, and now he was going to start over with the F4i. It would be easier, for sure, because we had the previous year's setup to work from. Also, the F4i was just an easy animal to tame in comparison. We had it well trained by that point.

"So what do we tell Honda?" I asked, "I mean I agree that we should use the F4i, I don't think anybody here thinks otherwise." I looked around for any objections, and then continued, "But they did give us two engines as a sponsorship, we owe them something."

Kyle replied, "Tell them that the car will be faster and better with the F4i, they want their name on the fast car, too."

Bobby added, "Yea, it's not like a Yamaha, we're going to another Honda engine."

"Yea, but I mean, should we just... ask them if we can sell their engines they gave us?" I asked.

"I don't see why not," Bobby replied.

I shrugged my shoulders "Okay. Are you going to take care of that?" I asked Bobby. One thing I learned in FSAE is that you never decide that the team will do something without making sure someone is responsible for it.

"Yeah," he said assuredly.

"Okay, let me know what they say and I'll put them on eBay as soon as I know it's okay."

Our discussion went relatively smoothly for such a big design change, but that was probably because we were all expecting it. We had the 600RR engines for two years and for two years we had put off designing them into the car. It was almost forced this time, and it just didn't work. Plus we had all seen Andy in the computer lab and had listened to the problems he was having. We probably should have made the decision a month before, but it was hard to give up two brand new shiny engines, still in their boxes.

A minor speed bump in the road.

The next time I saw Andy he was back in the computer lab and well on his way to the new design. Dave was gone, likely getting a short power nap in the library catacombs or a K-Mart parking lot. He would inevitably be back soon, unannounced and at random, to help Andy with the design.

"Dave is graduating in December, right?" I asked, without specifically directing the question to anyone.

"Yep," Bobby said, his tone clearly conveying that he had the same worry that I had: we wouldn't have Superdave in a few months.

"Is he sticking around for the competitions by any chance?"

I asked.

No response.

"I bet we can't pin him down to a yes or no answer, either," I added.

Trey replied, "I think he's planning to stick around unless he gets a job."

There seemed no point in worrying about it. Dave was a great designer and the design was mostly done. We could manufacture and compete without him. My biggest concern was that he wouldn't be at the competitions for the design event. His knowledge about racecars and design was a huge advantage in that event.

Trey, our former Team Captain and token grad student, seemed to shrug it off entirely, a departure from the approach he would have taken a year before. For the past three years, he had been one of the five or so people spending 50 hours a week on the team, but for the 2007 car, he was taking a much less involved role and wasn't letting anything get to him. I suppose that after spending tens of thousands of hours stressing out and dealing with never-ending intractable obstacles, you either relax or go completely insane.

I changed the subject, in part to force myself to not worry about it. "Did you guys get that DMC homework done?" I asked.

"I've got better things to do than that homework," Chris barked indignantly.

I gave a hushed laugh and said, "We've all got better things to do. Unfortunately they don't give degrees for formula work."

"They should," Kyle said, "I learn more here than I do in class."

This was true, without a doubt. Yoggi Berra once said "In theory, theory and practice are the same. In practice, they're not." School teaches you theory. FSAE is the practice that shows you how inadequate all the theory is.

Even within the realm of theory we were still years ahead of class. The homework in question was an introduction to Finite Element Analysis that basically entailed drawing a simple beam in a computer program, applying a load, and comparing the results with estimates from equations. As we were sitting there, Chris was using a much more complex Finite Element Analysis software suite to find the stress and deflection of the entire suspension assembly.

Nonetheless, he wouldn't get an A for that.

It looked like a reasonably easy homework assignment, so I figured I would bust it out the next morning before class.

"I still have to re-do Sommer's report for tomorrow," Kyle said.

Chris snapped, "I absolutely won't do that. That is a complete waste of my time. He got my original report and he can be happy with that."

Dr. Sommer, our faculty advisor, had assigned all the core team members to write a presentation on their system. Specifically, what their goals were and how they planned to meet them. After the deadline had passed and less than half the presentations were turned in, Sommer started to get upset. I had tried to make some sort of bridge between Sommer's reports and everyone else's lack of caring by asking the students personally, and explaining to Sommer that everyone was very busy.

I could see where Sommer was coming from and what he was trying to accomplish; by taking the time to present something we would be better prepared for the design event and the other members would have a better understanding of what we were doing on a more focused level. I also understood why some team members were reluctant to put a lot of effort into it. We didn't have much free time, and we would rather focus on the big things that would help us win. After a few days, everyone had submitted their presentations, most of which were half assed attempts at getting Sommer and I off their backs. Chris felt the exercise was entirely useless and made no attempt to hide his feelings.

"I don't have time for Sommer's fantasies about how this is going to help us," he said.

I gave a mediocre effort to try and get Chris to see things from Sommer's perspective, "Well, I think he's trying to get us all to think about the decisions we're making and get some insight on—"

"I don't need to think about my decisions, I made them and I know why I made them. If he wants to know, he can read my design report at the end of the year."

And so it went on. After he had a chance to look through the presentations, Dr. Sommer sent an e-mail to Bobby and me saying he was disappointed with them.

I was not surprised, and not happy that I had to continue to deal with the problem.

I reluctantly sent individual e-mails to everyone discussing their specific slides and asking them to add the things that I thought they were missing. I was trying to balance between keeping our advisor happy and not getting everyone pissed off at me by bugging them about the presentations. A couple days later, Dr. Sommer sent out another e-mail asking to meet with everyone the next day to discuss their slides because he was "not happy with what I see." He asked Bobby and me to sit in on the meetings. The first couple of discussions went okay, but then Ralph showed up and it all started to go downhill. Ralph walked in with no presentation, eating a bag of chips and generally acting like he didn't give a shit.

"I haven't got around to finishing the slides. I've been busy with other classes," he said.

"Do you know when you could have them done?" Sommer asked. Judging by his tone, I think he expected the same response from Ralph that I was expecting.

Ralph replied, "I think I may need to quit the team, I have too much this semester and I am not sure I can meet all the deadlines. I don't want to hold the team back."

In other words, he didn't realize it would take up as much of his time as we told him it would. Dr. Sommer paced through the social norms associated with those situations, saying stuff like "Well I'm disappointed, but I understand." I impatiently waited for him to leave so he wouldn't waste any more of our time. We had things to do.

After he left and closed the door, I said quietly to Bobby, "That's what we get for depending on a guy named Ralph." Bobby shrugged his shoulders without a laugh. I knew he was thinking what I was thinking; neither surprised nor upset. Ralph showed up at only a few meetings, and at none of our tests. I honestly did not see the team doing any worse or better without him. We handed him a relatively important part of the design and we would have to pick that up ourselves, but we were only left back where we had started: too much work and not enough time.

The next meeting was later that day with Chris. Five minutes after the meeting was supposed to start, Dr. Sommer came into the computer lab and asked Chris to join the meeting. Chris, never looking up from his screen, refused saying he was too busy. The two started to argue, and I just left the room. I knew Chris wouldn't show, Sommer would be pissed, and we would move on to the next thing. Bobby and I went to Sommer's office to wait for him to show up and say as much. After about five minutes Sommer walked in looking flustered and defeated.

He said, "I can't deal with Chris. He is not doing any of the things I have asked. I cannot work with a system leader that does not work with and communicate with me."

He paused for a moment before adding, "I don't think he should be a system leader."

Sommer was taking it personally, and I tried to diffuse the situation. "Chris is just that way sometimes; you just have to deal with it," Sommer started to interrupt but I kept talking, "Look, we can't do this without Chris, he has technical expertise that no one else has and on top of that we don't have anyone that can fill his position."

Sommer replied, "I know he's good technically, he's very competent technically, but he's not a team player."

We discussed Chris back and forth for a few minutes. It was obvious what Sommer was asking for, and we couldn't give it to him. I flat out said that removing Chris from his position was out of the question.

"No, we can't, we can't do it without him. We're already done with the design and it's his system. We all spend all of our free time on this car, and so does he. We can't stretch the team any further."

Then Sommer dropped the bomb, "I can't work with the team if there is going to be such lack of effort and disrespect given to my assignments and requests. I'm sorry but I can no longer be the faculty advisor for the team."

You gotta be shitting me, I thought. *You have got to be shitting me.* Bobby and I sat quietly stunned for a moment.

"What if you just meet with me and Bobby on Mondays and we will manage the team. You could talk to us and we could talk to the team," I suggested.

Bobby added, "Yea, we could do a military style chain of command."

Sommer shook his head.

I made a few more suggestions, trying to find a way to fix it so that we could have both Sommer and Chris still on the team, but it was clear that we were fighting a losing battle. Bobby said almost nothing during the meeting. Sommer was leaving and there was nothing we could do about it. Bobby and I, along with a half dozen others had spent the better part of three years building the team and now our advisor, our connection to the university, was leaving in our best year. We left the meeting and went directly to the bar. We had decided that we would not discuss it for a couple days, and we would let Sommer announce his decision.

Chris, who had lit the fuse, asked us later on that night, "So what's the verdict?"

"We'll talk about it in a couple days," I responded.

"What's the gist of it?" he asked.

"We'll talk about it in a couple days," replied Bobby.

I didn't have many categorical rules in life, but not quitting something partway through was one of them, especially when other people are depending on you. I had a lot of respect for Dr. Sommer before that day. I had worked for him, enrolled in several of his classes, and was always happy to complete whatever presentations or assignments he requested of me. In a matter of moments, I had lost more respect for him than I had for most other professors to begin with. The best teams have very strong faculty involvement and we now had none. I wasn't sure if he had destroyed our chances of winning, but he had certainly diminished them.

The properties of engineering materials are defined by how they react to extreme forces: ultimate tensile strength, hardness, melting point, etc. I believe people's characters are also defined by how they react when they are pushed to the limit. On an FSAE team, you get a lot of opportunities to see who people really are; we spent a lot of time stressed out, tired, and broke. Bobby was indignant, Ricky was standoffish, I was petty, and Chris was an asshole.

But we weren't fucking quitters.

4

Staying On Schedule

"No one wants to quit when he's losing and no one wants to quit when he's winning." – Richard Petty

Two weeks later we were still without a faculty advisor. Mitch Burris was one of our advisors but he was only an adjunct instructor, a part time faculty member. He would help us any time he could, but he wasn't a full time professor so he didn't meet the requirements for the school or the competition. The Director of the School of Aerospace and Mechanical Engineering had not officially approved Dr. Sommer's stepping down, but it was done. There was a sort of spat going on between them, one that seemed to repeatedly make its way into my e-mail inbox via the carbon copy function.

The Director was not happy with Sommer's "unilateral" decision to quit the team, and Sommer was arguing that his involvement was continuing in the form of allowing us to use his computer lab.

The problems associated with our lack of a faculty advisor were started to emerge. We needed an office and I was afraid the computer lab wouldn't be there after another month or so. We also needed an advisor that had some pull in the department in order to ensure that we had other resources.

In 2004, we were kicked out of our shop because a professor needed space for his research. Without a professor to push for us, we could end up having to build the car in somebody's parents' garage. It was possible that we could have lost our shop just because some tenured professor needed space to test the structural integrity of dog shit.

Our team was not going to last very long without a faculty member to ensure our progress, or at the very least, our survival. It was upsetting that we had to deal with this issue on top of fundraising, managing, designing, building, and testing. I wanted to focus more on the car and let Bobby deal with future team issues, but I knew that a new advisor was necessary for the current car as well as future years. Chris stayed with the team, happy to have Sommer gone, but also clearly concerned about our situation.

Our meetings continued as usual, and I tried to focus on the car design, and perhaps more importantly, how the hell we were going to get it built and tested in time. Chris had proven my belief that he worked too hard to let him go. Even still, he had more work to do than he had time for.

"Chris, when are you going to machine the wheel centers?" I asked, going around the room and trying to keep the schedule updated.

"I don't have time," He shot back defiantly.

"You... do you have time to code them or do you just need someone to babysit the machine?" I replied, trying to be diplomatic.

"The coding is done, I just don't have time to sit in front of the CNC all night while they cut."

"If we can get together and start on one of them after the meeting, I can machine them."

Chris shrugged. I took that as a yes and added that day's date to the completed wheels entry on my spreadsheet. Two days behind the original date, but good enough.

CNC machining was a dreadfully boring process if done right. Programming was the hard part, and Chris was an excellent programmer. The actual machining was just time consuming. You would sit in front of the machine and watch it cut metal, making sure it didn't do anything you didn't intend for it to do. Of course, occasionally you would put a clamp in just the wrong place and the tool would slam into it, exploding into a thousand hardened steel pieces and sending the machine into alarm, flashy red light, everyone-look-at-the-guy-who-fucked-up-the-machine mode. It was exciting. And expensive. I expected none of that.

"Do you have the intake design done?" I asked.

Chris answered quietly, "No, we need to find out if we can have the intake extend past the back of the frame."

"Okay," I said, waiting for him to finish.

"In other words, you need to send in a rules clarification," He responded.

"I don't understand the clarification needed. The rules say it needs to be within the roll structure. If it's—"

Kyle interrupted, "But we don't know if behind the frame is within the roll structure."

"They define the roll plane as the top of the hoop and the rear wheels. I don't see why it wouldn't be legal as long as it is within the rear wheels," I answered with a little less confidence than I should have.

Chris said, "It's your job to clarify the rules."

"Look, if you don't think your design will pass the rules, you ask for the clarification. I'm not going to clarify every rule when I think it's obvious."

Bobby added, "It's not obvious."

I was a bit confused as to why it was my job to clarify rules with the rules committee regarding other people's designs. It didn't seem efficient or reasonable to me. Mostly, I felt like I was being attacked by half the team, so I went defensive and changed the subject.

"I'm not going to send in clarifications that I think don't have any merit. If you want to send it in, be my guest," I turned to Phil and continued, "Phil, how is the fuel system coming?"

"We're still waiting on the seat location to finalize the tank size," He replied.

Kyle jumped in, "The seat is finalized. It's been done for a week."

"Really?" Phil asked.

"Yea," I said, "We need that computer model soon; the electronics depends on the fuel tank location."

"Okay, I'll get it done."

"This week?" I asked.

"Maybe next week, I've got a lot—"

I interrupted, "We really need it this week; its deadline was a week ago."

"I'll try," He said.

I hate "I'll try." In FSAE, more than anything, it means "I won't." It's an apology for a failure that hasn't even happened yet.

"Is it okay if I go ahead and model the tank part so we can start to work around it?" I asked.

He shrugged his shoulders. This would have been a good time to balance deadlines with concern for teammates' individual schedules and feelings about their part on the team. But I hated missing deadlines.

"Anyway, you have to finish the cost report, and there is no 'I'll try' with the cost report, it has to be done, and it needs to be done at least a week early," I said authoritatively.

"Okay, it'll be done two weeks early," Phil said with his usual optimism.

"Alright, I'll send you the fuel tank model when I'm done and we can get it made. There's no reason to wait on manufacturing it. Speaking of manufacturing, for all the new members…"

There was a small handful left at this point. We had whittled them down in no time. "…you need to be working on the car." I looked at one of the familiar faces and said, "I know you've been working with Chris, what are you working on exactly?"

"Umm, we've been doing some dyno testing and—"

Chris interrupted, "Wes is working hard; we've got stuff for him to do."

I thought, "Wes, that's his name. Remember Wes."

I would have liked to know what Wes was working on, but Chris was being defensive about his system. Defensive in the form of not letting me know what was going on.

I shrugged it off and continued, "…Good enough. Brett, are you working on anything?"

"Umm, I've been helping Bobby make some frame pieces."

Brett was the other new member who stuck around. He was a quiet guy, but he spoke with confidence. I had seen him around a lot but hadn't talked to him much. He spent a lot of time with Bobby working on the frame. I could tell he hadn't shaved in a few days, and his shirt had grease stains on it, which meant he was already pretty much a full blown team member. He had a girlfriend, which was good; in FSAE it's hard to keep an existing relationship going, but it's nearly impossible to start a new one.

"Sounds good." I said, "Ricky, what are you up to?"

"PR stuff," He replied.

"What PR stuff?" I asked.

"Raising money," He answered, "You know, so we can pay for all this stuff."

I was on the car side of things, not the logistics side, so I could be forgiven for forgetting about raising the $40,000 needed to build the car and get to the competitions.

"Lauren and I are also trying to get some media coverage in the paper or possibly the local TV news. It is looking promising."

Ricky had recruited Lauren to the PR team. Lauren was a journalism major who worked in the Aerospace and Mechanical

Engineering office. We were always in there spending money or picking up packages, so she knew us pretty well.

"Sweet." I said, "Kyle, how are the brakes coming along?"

"Huh?" he asked, and then answered, "Everything is modeled in the computer, and Chris and I machined the pivot and I bought $140 worth of tiny little bearings."

"Is the design finalized?" I asked.

"It's done. I called it good," He replied confidently.

"Calling it good" was our team's way of saying "trust me." It worked out really well; no one would call anything good without being sure it was good. And so we could save a lot of time by just trusting the call. We had rules, only senior team members could call it good, and on things that were super critical we would always get at least two people to call it good.

"Okay, so…" I looked through the spreadsheet schedule. "Bearings purchased one week ahead of schedule, design finalized on time. Looks like someone is getting a gold star," I said with goofy enthusiasm.

I noticed Dave in the corner of the room. Dave didn't have any official responsibilities or deadlines, but I knew he would be working on a few things, so I took the opportunity to see if I could get some deadlines out of him.

"Dave, are you working on the steering rack this year?"

He replied quietly, "The design is the same as last year. It only needs to be manufactured."

"Okay," I said quietly, "are you going to manufacture it?"

He shrugged his shoulders.

"Alright, well I'll work on the clamps like I did last year. Chris, can you find some time in the next few weeks to bust out the extension tubes?"

He shrugged his shoulders. It's hard to keep a team on track with everyone being standoffish and it was getting irritating. It was especially hard to pin down manufacturing dates with Dave and Chris. Neither of them had any respect for what I was trying to do and would rarely give dates, so I just filled in estimated dates of completion for them. I think that pissed them off even more.

"Andy..." I continued with the other team members, giving gold stars where deserved and trying to make sure we weren't forgetting anything.

At the end of the meeting, I found Chris and we went downstairs to start on the wheel centers. The CNC mill was in the basement of the engineering building in our department's machine shop. It was a relatively small room built half underground with covered up windows and machines seemingly placed randomly. The walls were cinder blocks painted with that greyish green color you only see in high school bathrooms and sad movies about East Berlin.

Chris had made a jig to properly align the front and back sides of the wheel centers, so the whole thing was a streamlined process that took the technical knowledge of your average English major. Chris was taking the opportunity to give Bobby a short lesson on how to use the CNC machine, so that we could have more than just a few people who were able to run it. After it was set up and running, the two of them left and I sat there waiting for the machine to finish. Every 90 minutes or so I would flip the wheel, or add a new block of aluminum, press a few keys, and wait.

It seemed like there was starting to be a rift between the senior team members, like there were two teams forming: Kyle and Ricky on one side, and Chris and Bobby on the other. Most of the other members spent time with Chris and Bobby, probably because that was the group that would be around the next year. Trey seemed to seamlessly move from group to group, avoiding any affiliation. Dave and Phil each did their own thing, and I think I was just pissing everyone off. A cohesive team is often better than a split team, but we were getting things done so I was okay with the way we were organizing ourselves. Building the

61

car was just a series of small steps and everyone was working on their assigned parts. The steps would be done, and the path would be completed on schedule.

While waiting on the mill, I searched for racing videos on the internet. I found one that showed training for Formula 1 drivers. They were balancing on a large ball and catching tennis balls that people where throwing at them. Driving is all about balance, reaction time, and hand-eye coordination, so this seemed clever. I looked around the shop for similar items. I found the tennis ball, but no large ball so I improvised by balancing on a big piece of aluminum. And so for the next few hours I stood with one foot on the aluminum bar, like the Karate Kid, throwing a tennis ball at the wall and catching it.

After the last wheel finished, I started sweeping up the machine and putting the tools away. I was just about finished when the door opened and Billy, the shop manager walked in.

"You been here all night?" he asked.

I looked at the clock on the wall, "Dammit, is it already eight?"

"You formula guys need to get girlfriends or something, and stop spending all your time on the racecar."

"Where's the fun in that," I said with a smile, "I'll see you later, I'm outta here."

"Going home to sleep?"

"No, I got class in an hour; I'm gonna go take a nap."

I walked upstairs to the computer lab, balled up my coat on the desk, laid my head down and went to sleep.

Twenty minutes later I was startled back to the unforgiving world of FSAE by someone shouting in the computer lab.

"They're towing the trailer; can you get there before me with your motorcycle?"

I blinked and squinted to see Bobby holding his phone to his head and yelling in my direction.

"Www...what?" I asked.

"They're towing the trailer!" He repeated.

"At the north shop?"

"Yea."

"They're towing the...? Goddammit," I grabbed my coat and headed towards the shop. On my way I had plenty of time to wake up and get pissed the hell off at the fact that some asshole was towing a trailer in an industrial complex with plenty of street parking. I caught up with Phil just as he pulled into the shop. The tow truck was just leaving and the parking officer yelled at me and Phil.

"Get away from the tow truck or you will be arrested," he said authoritatively out the window of his Parking Authority Ford Escort. He got out of the car and walked towards us. He was wearing white shorts and a cheap hat with a city logo screen printed on the front. I wondered what kind of person, at some point between high school and the age of 40, decides that they want to spend their life doing a low paying civil service job where everyone hates you. I have never had any respect for people like that, only disdain that was usually passive, but not that day.

He continued, "This trailer has been par—"

I interrupted, "This is an industrial complex five miles from anything, there are always trucks parked around here..." he started talking, trying to interrupt me but I continued, "...and none of them are ever in the way!"

He responded more calmly, "It is illegal to park a vehicle in the street for more than three days and someone complained so we—"

"Well I want to complain about that car," I started pointing at vehicles parked in the street, "and that one, and those two over there. That one has been there for a fucking week!"

"If you want to file a complaint with the city..." as he continued talking, my better judgment kicked in and I walked away. It was towed and there was nothing we could do about it. Phil got the information so he could take care of it later.

Two days and $200 later it was back in front of the shop, and we were looking for a new parking space before our three days expired. It was a pain in the ass getting it out of impound, because they wouldn't let us take the trailer without proof of ownership, but the trailer was owned by the university and we didn't have any papers to show that. Phil managed some deal, I didn't know what it was and I didn't care. Phil was not the best engineer on the team, but there were things he was good at doing and I liked to leave those things up to him. It was good to delegate responsibility sometimes; it meant I got 4 hours of sleep a night instead of no hours of sleep a night.

5

The Business of Racing

"The answers to your next three questions are: one hundred thirty, three hundred thousand, and no." – Dr. Bob Woods, University of Texas Arlington FSAE Faculty Advisor

Racecar Engineers really only have one job: win races.

It's not the Racecar Engineer's job to advertise, or to recruit. It's not even his job to follow the rules; cheating and winning still falls within the ultimate goal. Cheating and getting caught does not, however, because getting caught usually eliminates the possibility of winning. We wanted to cheat and win, but then we would take off our engineer hats, and put on our risk assessment hats and see that it wasn't worth the risk.

Unfortunately, unlike professional race teams, we didn't have one job. That is, we weren't just Racecar Engineers. We would also be managers, writers, artists, fundraisers, publicists, recruiters, and whatever else we needed to do so that we had the resources to be Racecar Engineers. It was more like running a small business.

For the life of the Sooner Racing Team, we had been relatively unknown. Most of the time, when mentioning the University of Oklahoma's racecar, people would assume we were talking about the solar powered car. The solar car got a lot of publicity for whatever reason, and we got very little. We had been trying to fix that by bringing the car out to show off at OU's real attraction: football games. We would park the car in front of the engineering building, Felgar Hall, and wait for people to stop by, ask the same three questions, and hopefully buy a t-shirt.

The first question was always the same: how fast does it go. We hated this question. People wanted to know how fast because they assume that top speed is a good measure of the performance of a racecar. If the racecar is built to break some land speed record at the Bonneville Salt Flats, then top speed is important. Otherwise, it is trivial. In fact, in our case there were good reasons to artificially limit the top speed: mostly to make the best use of the transmission gears available. The answer we gave to this question was 100 miles per hour. In fact, we didn't care how fast our cars went because before we got close to the top speed, we had to slow down for a corner.

The second and third questions that people had were usually the same, though not always in the same order. Some people wanted to know how much it cost, and then if they could drive it. Some wanted to know if they could drive it, and then how much it cost. The answers were no and $16,000, or $16,000 and no, or sometimes "a lot" and "not a chance." A few people wanted to know what engine we used, and a few drunken rednecks would just yell "WOOOO GO NASCAR YEEAAH!"

$16,000 was not strictly accurate. A small percentage of the total points in the competition were based on how much the car would cost to make in production. The cheapest car got 30 points, the most expensive got none. We, like most teams, usually came in around $16,000. Every team had to submit a cost report each year, detailing the cost of every component on the car, or at least what the cost would be if they were making several hundred of them a year. Teams could get clever and list the cost of something as if they made 300 of them at once using a badass two million dollar CNC mill even though they only made one of them in their garage with a hammer and a rusty screwdriver. This was debatably fair and legitimate, because there were some teams that really did have badass two million dollar CNC mills. It didn't really matter anyway because cost was only 3% of the points, most of the good teams were all around the same cost, and all the teams would mingle in the grey area of the intent of the cost event. The teams that didn't, well they just weren't trying hard enough. Exploiting the grey area of the rules is as much a part of auto racing as anything else. In cost, it was just really easy to do, because there was so much grey area. Also, if the cost judges caught a team flat out lying, they would just add 150% of the cost of the item in question. So even with our risk assessment hats on, we usually pushed the limit a little and accepted a few 150% hits.

$16,000 was also inaccurate because it didn't account for all the time we spent designing, testing, and refining the car. If you were to commission a group of 15 college interns to design and build one of these cars, you'd be looking at around $300,000. Of course, all of this requires explanation, so we just said $16,000 and left it at that. Most people didn't want an explanation, they just wanted a big number; same with the "how fast" question. They asked, we told, and they walked away disappointed. The University of Texas at Arlington answered the "how much" question with the $300,000 answer. We probably should have gone that route, if only to give one impressive number.

It's really sad that everyone asked about the least impressive stats of the car. No one wanted to know that it could hit 3g's in some corners, or that it could accelerate from 0 to 60 MPH quicker than most cars that actually cost $300,000. It was a racecar, and all they saw was a hopped up go-kart. But they weren't racecar engineers, they were football fans. Drunken football fans.

There were no major sports teams in Oklahoma at the time, and OU had a great football program, so that was it. For at least four months every year, OU football was to Oklahoma what movies are to Hollywood. The atmosphere was impossible to ignore on game days. Some drunk guy half a mile away would yell "Boomer!" and everyone in audible range would yell "Sooner!" and we would repeat two more times, or more if the boomer guy was too drunk to count. It was just something OU people yelled. Oklahoma State had "Orange Power" which sounded like a bad energy drink. Texas had something that started with "Texas," but whenever they yelled it, all the Oklahoma fans would yell "Sucks!" and since I was always in a crowd of OU fans I never heard the official ending.

In any case, we were a University of Oklahoma team, and game day Saturdays were filled with OU pride so it was a good time for us to try to do the public relations thing.

Ricky was standing behind the t-shirt table, occasionally selling a shirt. Bobby, Kyle, and I were sitting in our folding chairs, waiting for people to ask questions.

"Hey Matt."

I turned around to see an old friend from freshman year.

"What's going on, man?" I asked, trying to remember his name. *John. Is it John? Yea, John, that sounds right.*

The group I hung out with my freshman year was the party-all-the-time crowd. Even back then I only hung out with them a couple times a week, trying to focus on school enough to keep my grades up. Most of them spent their first year too drunk to focus on anything, much less school. I was impressed that they had the energy to party for a solid year; I always felt like party houses lost their appeal after eighty consecutive parties. It was a good place to go if there was nothing else to do, but when you're on a formula team, that is rarely a problem. Like many of my pre-formula friends, most of them had fallen into I-can-barely-remember-your-name obscurity. It's part of the tradeoff.

He answered, "Just here to watch the game. What have you been up to?"

"You're lookin' at it," I said, pointing to the car, "This thing takes up pretty much all my time."

"Yea, I haven't seen you in a while. You should stop by the house sometime."

"Yea," I said, trying to sound like I was interested, "you still at that place off Boyd?"

"No we're over by Sarkys, in that apartment complex just to the east. Number 226, it's in the far back corner."

The marching band always paraded down the street in front of our spot shortly before the games. It was everyone's cue to get to the stadium. The drummers were banging out the first part of "Boomer Sooner" and the horns had just joined in. It was too loud to talk, and I took it as an opportunity to end the conversation.

"Yea, I'll come check it out sometime." I probably wouldn't and I knew it, but I liked to keep my options open.

"Take it easy," he said, walking away.

On the team, we always talked about how we were missing out on the "college experience." We were always working on the racecar, and didn't have time for… whatever it was that everyone else was doing. I had ventured out into the party world a few times. It was a scene of dozens of guys, all with the same spiked hair and the same popped collars on shirts just different enough so that they wouldn't be mistaken for uniformed wait staff. They were all making small talk with dozens of vapid, half drunk girls, and no one had anything interesting to say.

"So, you uh… come here often?"

"Ha ha Yea, I live here ha ha."

It was retarded. When it came to deciding between the racecar and the "college experience" with its empty people, cheap beer, and vomit covered sidewalks, it seemed like an obvious choice. I was always shocked when I explained that to people and they just looked at me dumbfounded. I wanted to jump up and down and yell "A racecar! It's a racecar! That's that coolest thing ever!"

But for some people, it might as well have been a giant toaster.

I looked over at the car and noticed another familiar face squinting carefully at the engine.

It was Wes, one of the few new members left.

"Hey Wes, how's it going?" I asked, walking towards him. "You here for the game?"

"No, I just came to see the car." He said, still looking at the car.

I think we might have a winner.

"So what do you guys do for the muffler, is it baffled or do you use packing?"

"We've tried several different ideas. A perforated tube with packing is the popular choice among the other teams and seems to work well. Penn State was saying they used a triangular inner section to get more surface area for the packing."

"Fiberglass?"

"Uhh, we use muffler packing from the motorcycle shop, it's probably just fiberglass. Any dense fiber that won't burn or melt will probably work the same. I put carbon fiber in the muffler I built for my Dodge Dart, just because I had a bunch lying around that was too dirty to use for anything else. Although it is loud as hell, so maybe that's a bad idea."

Wes looked up from the car and asked, "A Dart? What size engine?"

"400, big block. It's my project car."

"I thought it was a Plymouth," Bobby had walked up behind me to join the conversation.

"It's a... well no one knows what a Plymouth Valiant is, so I call it a Dodge Dart. Same thing. Honestly half the parts on it are from a Dart, the brakes, the rear end. The engine is from a Dodge Magnum. Fuel injectors are from a Mitsubishi Eclipse, the throttle cable is from a Honda motorcycle."

"So you just call it a Dart," Bobby said.

"I call it a bottomless money pit. It still needs wheels and tires, and a transmission rebuild. It's the physical incarnation of my poor money management skills."

The marching band had come and gone, taking with them most of the crowd. A few fans were left scattered around tailgate televisions. We decided it was time to pack it up. We pushed the car into the basement of the engineering building and locked it up. Bobby would tow it back to our shop the next day, after the 120,000 football fans were gone for the week.

"Matt, are you coming to watch the game at my parents' house?" Kyle asked.

"Nah, I think I'm gonna get some homework done," I said, partly lying. I would get some work done, but I would probably end up watching most of the game. "I would like to get some lunch, though, if you guys are up for it," I said, pointing to Wes and Bobby.

"What were you thinking?" Bobby asked.

"I dunno, Louie's? Somewhere with a TV so we can watch the game."

"Louie's sounds good." Bobby said.

"You in?" I asked Wes.

"I'll go with you, but I can't get any food. I don't have any money and I have food credits that I need to use."

The University had a policy of packaging food plans with dorm leases, and they had a requirement that freshman students live in the dorms. It was another one of the ways they tried to baby-sit the students. I figure if you've made it to college and haven't yet figured out how to manage your food budget, it's time to learn. For some students that may mean standing on the street corner at the end of the semester with a sign reading "Starving student, please feed," but of course that is a lesson well learned and not soon repeated.

We made our way to the restaurant. A waitress instructed us to stay at the front while she had the busboy clean off one of the tables. Most of the pre-game crowd had shuffled out, leaving the tables covered in empty beers and half eaten pizzas.

"So where are you from?" I asked Wes.

"Fort Worth."

I nodded and asked, "So why Oklahoma?"

"Oklahoma is a good school and it's easy to get into. It's also pretty cheap."

"Yea, I've met a lot of people from Texas here." I thought for a moment, "Probably about half the people I know here are from Texas."

A bubbly blond waitress picked up three menu's and said, "Follow me you guys!" with a smile. It was no secret that OU football meant really good tips. This attractive lady and her cleavage exhibition were likely making a small fortune. No wonder she was so happy.

"I wanted to go to Texas A&M, but if you're not in the top two percent of your high school class, you might as well not even try."

"OU is definitely not difficult to get into," I said, looking down at the menu.

We sat down around half the table so we could see the TV hanging on the wall.

"Do you have a job or a tough schedule this semester?" I asked, though I might as well just have directly asked "How many hours can we expect you to work on the car?"

"No job. The classes seem pretty easy, but I am taking sixteen hours."

"Yea, I recommend loading up on classes early on, then gliding through your last year without problems. If you need a foreign language, I highly recommend Cherokee."

"Cherokee? Like the Native American language?" He asked.

I smiled and nodded, "Ricky and Kyle are taking it too. It's like kindergarten. It's the easiest class I've ever had. Bobby is taking it next semester just to boost his GPA."

"So can you speak it?" He asked.

"Heh, no. I think I could probably say like fifteen things. I forget stuff right after I take the test over it. I know Wodige means Brown, and I think Saloli is Squirrel. I've gotta take Cherokee two next semester which might be harder, more like first grade. Actually, next semester I only need eleven hours to graduate, but I need twelve to be full time and get my student loans, so I'm taking Tennis. It's a one credit hour pass/fail class. It's gonna be a sweet semester."

Wes brought the conversation back to the car again.

"So what is the most important part of these cars? I'm guessing the suspension?"

Bobby answered, "Tires. And suspension that makes the best use of the tires."

I added, "All the performance has to be manifested through the tires, because they're the only control surface. Also, I think driver and aero are two things that are waaaay undervalued in FSAE. It's hard to say how much; we don't have a lot of data on aero and it's hard to quantify the driver since we don't know how good other teams' drivers would be in similar cars."

Wes asked, "Are the competitions pretty intense, or do most teams have everything fairly well sorted by that time?"

Bobby and I laughed.

Asking a Formula SAE student about a competition is like asking a Vietnam vet about the war. They just look out the window with that thousand yard stare. "I've seen some things man... I've seen some things." And then they proceed to explain the event to you in terms that normal civilians wouldn't understand.

"We were coming in from the south, four clicks from tech inspection, when the VC (university of Virginia, Chesapeake) came in from all sides. We were surrounded, there were sounds of engine backfires everywhere. The Brazilians were dragging their dead car back to base camp; the front wheels were gone, man, ripped off when

the uprights buckled under the moment reaction in brake tech. "Where's the ARB!" Kyle yelled into the radio "We can't properly distribute the roll couple without the rear ARB!'"

We ate our lunch and talked about cars and racing, and some about school. Wes seemed like an enthusiastic person who could help the team. He knew a lot of random facts, like he spent his weekends surfing Wikipedia.

"Now we get to play my favorite game," I said as the waitress approached our table with the check.

"What's that?" Bobby asked.

I responded, "I call it 'Will Matt's check card be declined.'"

I always lived so close to the edge of my budget that, from time to time, my checking account would be empty at inopportune times. I could have balanced my bank account, but that would have taken all the fun out of my game. Also, it would have been really depressing.

"Do you usually win this game?" Wes asked.

"You never win," I said handing my card to the waitress, "you just get lucky enough not to lose."

6

Falling Behind Schedule

"The cost of racing hasn't increased in thirty years. Back then, it took everything you had. And it still does." – Unknown

I averaged about 50 hours a week working on the race team. I also went to school full time. That's supposed to be 90 hours a week for those of you keeping track. To be honest, I rarely spent more than 20 hours a week on school, so we'll call it a 70 hour work week. With all of that time taken up, I still managed to run, ride my mountain bike, work on my own car, and catch an occasional episode of The Simpsons. I could have probably even fit a flimsy relationship in there if I managed my time. It's not that difficult. 70 hours a week is 10 hours a day. Add 7 hours for sleep, and you get 7 hours a day free time. Even if you blow an hour for lunch and an hour for dinner, you still have time to run a few miles, ride some mountain bike trails, change an alternator, or watch a movie. That's every day of the week, and I assure you I didn't change my alternator every day.

There is always time. Even people who are very busy can generally spare several hours a week for something they want to do. Especially college students, no matter how high the GPA or how many hours the part time (or even full time) job. There are exceptions, but they are few, and those people are usually extremely successful. Unfortunately, everybody thinks they are the exception. All the time, I had people tell me that they couldn't help with the team because they were too busy. Every time I saw someone that used to be on the team, I would comment that I hadn't seen them around. I always got the same response.

"Yea, I want to, but I'm just so busy. I've been working 20 hours a week and doing school and it's just killing me."

I always responded politely with something like, "I know how that goes. Well, if you ever get time, stop by for one of our meetings," and I walked away disappointed and with an image in my head of them sitting on the couch watching The Matrix, smoking a doobie, saying, "Man, all this working is killing me. I need a break." I tried not to be indignant, but I was sick of hearing the excuses.

Phil was falling behind schedule on several things, and by that point in the year, my indignant days were beginning to outnumber my empathetic days.

Usually, I would stick up for Phil. He had his good points and had accomplished some good things in the past. I would always point this out, but it was getting harder and harder to mention the fine job he did two years ago on the jig table. Phil had been dodging his duties since the beginning of the school year. He was in charge of the brake dynamometer, the team shop, the fuel system, and the cost report. By the time November came around, he had done half of what he was supposed to on the brake dyno, and none of the fuel system. Some people complained behind his back, and my opportunity to set an example had deteriorated into my own complaining while he wasn't there. It must have gotten back to him, as those things usually do, because Phil approached me one day in the machine shop and said, "I owe you an apology."

"What for?" I asked.

"I haven't done anything with the team lately. But I've just been so busy with the…"

I let him finish, though I wasn't really paying attention. I read in some business management book that it's a good idea, as a manager, to let people save face. So I always let people tell me their excuses even though I didn't want to hear them; especially this one. Every other time I saw Phil in the previous couple months, he was working on something other than the car.

Later that week I was exhausted, sleep deprived, and at the end of my sympathies. Someone made a comment that the jig table was the only thing Phil had ever given the team, and I agreed. "We should fire him," I said, followed by a brief silence.

"Can we do that?" Ricky asked, seemingly shocked that I had suggested it.

"Yea, we fire someone every year. Last year it was Caroline, the year before that it was Tim."

"Well, if he doesn't do the cost report, that's a fireable offense. I don't think we should fire him now, though," Ricky said.

Ricky was defending Phil, and while they were certainly not enemies, it seemed like a signal for me to back off the firing thing. Bobby said he would talk to Phil, and tell him that the cost report was the only thing keeping him on the Detroit trip. It was not car related, so I let Bobby handle it.

"Plus, most people have been busy with exams," Kyle added. "Now that those are over we can spend more time on the car."

I tended to forget that most of the team put more effort into their grades than I did. I was usually smart enough to skim by with a B- with almost no effort. I was a little worried about my statistical probability class; I hadn't done well on the previous exam due to a complete lack of studying. I would have calculated the statistical probability that I would pass, but I had no idea how to do that.

In any case, I had to get back to the wing design which was almost behind schedule. I couldn't expect people to keep their schedules if I wasn't keeping mine. And I certainly couldn't help them if I was also behind.

The next morning, Bobby and I arrived at school early, on a mission to accomplish a task too important to be left to one man. For the previous two years, our team had almost missed registration for the Detroit competition. Registration opened at 9am central time, and all the teams that wanted to compete in Detroit all registered within an increasingly short time span until all 130 spots were taken. In 2005 and 2006 the competition filled in just a few hours and our Team Captain almost didn't make it. So for 2007, both Bobby and I showed up at the office ready for registration, with plenty of time and more than a little determination.

At 9am we were on the website, and we had basically kidnapped Vicki, the office assistant who had the department's credit card. For a few minutes, we were trying to find the link for registration, which was getting difficult since the website was taking so long to load. Apparently, the web page server couldn't handle the load of would-be registrants all trying to register at once. Each page took several minutes to load on Bobby's computer. I was on another computer across from him also looking for the registration page, but my connection wasn't any faster so it wasn't helping. After about ten minutes, he had found the registration page and was waiting for it to load. I pointed my web browser at the list of registered teams. In less than 15 minutes it had made it up to 75. The registration page loaded and Vicki quickly entered the credit card information. Bobby clicked the submit button and we waited for another three or four minutes for the page to load. While we were waiting, Dr. Woods, the Faculty advisor for the University of Texas Arlington FSAE team called and said, "I see a hundred and sixteen teams registered and you're not on the list!" Bobby told him we were in the middle of frantically trying to get registered. The webpage finally loaded: "Credit card information invalid." Vicki re-entered the information, quickly but a little more carefully this time and clicked the submit button. We were hoping it was a problem entering the information and not the credit card itself. While waiting for the page to load, I refreshed the list: 130 registered teams. Registration had only been opened for 17 minutes, and it was full.

"A hundred and thirty teams," I said dejectedly, "we missed it." I sat back and sighed.

"Maybe not, I clicked the submit button," Bobby said optimistically, "it just hasn't loaded. Check the list for our name."

I scanned the list. Sure enough, number 125 – University of Oklahoma. A few seconds later the credit card approval page loaded on Bobby's computer: "Thank you for Registering!"

"Holy shit that was close," Bobby said.

Using the engineering department's credit card was not the best solution but we didn't expect the event to fill up that fast. We would have to figure something else out for the California event registration which would open for us a month later. In any case, neither Bobby nor I had $500 in our checking accounts to loan to the team. It was near the end of the semester and I probably couldn't have fronted a $50 registration fee.

Being completely broke is an important part of any college curriculum. When you have no money, you learn lessons that no professor can teach you, like how to sell all your stuff on eBay to pay rent, or how to do three loads of laundry with only two dollars in quarters and an all-night car wash. Problem solving skills for the real world. Part of my flat broke story was my history of cars. As a "Car Guy," I think people expected me to drive a custom hot rod or a slammed Honda with stupid looking stickers on the side. Instead, I opted for the barely running shitbox. Actually, that should be shitboxes, since there were several of them.

The first was a 1979 Plymouth Volare that I bought for $600 of my hard earned burger flippin' money. I was the first of most of my friends with a car, so we drove it everywhere. I've seen several lists of the "worst cars ever made" variety, and the Volare was on about half of them. I had pretty good luck with mine; whenever it did break I would go to the local junkyard, find the pile of old Plymouths and get my part. I knew nothing about cars when I first got it, and by the time it left I could change water pumps with the best of them. That car was the reason I am a car guy. You don't spend $800 at the transmission shop to get your $600 car fixed, so when it broke I would have to figure out how to fix it myself. Before too long I was fixing things that weren't broken, and that's really all a car guy is.

The one that came immediately after the Volare was my Neon, the only car I've ever owned that was even remotely acceptable as an actual car. Only remotely, however. It made it through the first two years of college with a lot of replaced parts and one engine rebuild. Later it started to have more major problems. I lost interest and stopped changing the oil about twenty thousand miles before I sold it. I took off some of the more valuable parts and sold them on eBay, followed by the soulless shell of a car which pulled in a whopping $38. Around that time, a co-worker of mine had bought a new car and gave me his beat up Ford Explorer just to get rid of it. I fixed it into running condition and proceeded to deliver pizzas in it, putting 100 miles a night on a car that someone else didn't even think was worth selling, much less driving. The clutch went out halfway through a delivery, and I sold it on eBay a week later for a couple hundred bucks. Then for a while I went without a car, riding only my motorcycle. There were times when I would bundle up with four layers of clothing to stay warm. I rode around looking like the Michelin Man on a motorcycle in 15 degree weather.

When I got tired of riding in the freezing rain, I borrowed my dad's F-100 pickup truck. My dad had stopped driving it since buying a new Jeep, so it was just sitting in his driveway. It got the job done, but that was about all you could say for it. It got 60 miles to the gallon, but that was just the oil. Every other time I drove it, I would need to put a quart of oil in it. I threw the empty quarts in the bed of the truck and had amassed quite a collection. Some of the oil burned up in the engine creating a stream of smoke out the exhaust like an acrobatic airplane. Most of it leaked out from a dozen cracks and holes in the engine making the engine bay look like a tar pit. Up until about halfway through the school year it hadn't yet completely died on me, which is all I really ask of a car. Unfortunately, that was, again, too much to ask.

"So I was driving the truck back from my parents' house last night," I said, talking to the team members in the shop who were gathered around the frame table. "when I noticed the oil pressure gauge drop to zero. I pulled off at the next exit and into the Wal-Mart parking lot to get some oil. I walked back to the car section and got a gallon of oil, and then wandered around Wal-Mart for a while, letting the truck cool down because it was running a little hot." I had the tone of voice like I had an interesting story, so a small crowd had gathered around to listen. "Over the intercom, someone said, 'Would the owner of a white Ford F-150 please come to the food entrance.'"

Bobby started to laugh and said, "uh oh."

I continued, "I didn't think too much about it, because I drive a white Ford F-100, not an F-150. But it was about time to leave anyway so I started to head out. As I got closer to the front of the store I started to smell smoke, and I figured if anyone's truck is on fire, it's probably mine. So I walked out the food entrance and sure enough, there's my truck, on fire."

"Ha!" Kyle said with a smile.

"And not just a little on fire, it was *on fire*," I added, with my arms outstretched as if to show the size of the fire.

"No way..." Bobby said.

"Freakin' way, man. Right there in the Wal-Mart parking lot."

"What... what did they do? Did the fire department show up to put it out?" Kyle asked.

"Yea, they were just pulling up as I came out. They didn't

even ask how it started; they just saw the pile of empty oil containers in the bed and drew their own conclusion. After they put it out, the police asked to see my insurance, so I pried open the half melted glove box and handed him the charred black and soaked insurance card, where you could barely read "Allstate" in the upper left corner. He said, 'You can just call me with that info in the morning.'"

"Wow," Chris said, "That sucks."

"Yeah, I know," I said nodding, "I was there."

I answered some questions about the fire, and where the truck ended up. It made for a good story, for sure, but I needed a vehicle more than a story. But that was that and I was back to being without a car.

The upside was that I would save money on gas and oil, which was actually a very good thing because I had lost my favorite debit card game a couple of times in the previous month and things weren't looking any better. It's surprising how close you can get to the financial edge of a survivable living situation. Every time I thought I was at the edge, I came a little bit closer. Something like a speeding ticket would have put me well into the red and then I would have actually been living in the race team shop instead of just "living" there.

7

The Occasional Collision

"The price for men in motion is the occasional collision." –
Carroll Smith

Ever since I had become Chief Engineer, Chris's attitude towards me had been almost always condescending. My initial reaction was usually frustrated anger, followed by personal reflection, and eventually I would just brush it off. I felt like he just kept taking jabs at me and responding to every statement I made with a condescending and personal attack. After a few months I just got tired of it.

I had updated the 3D design software in our shop to the newest version and sent out an email letting everyone know. The new software was slightly different, but it was faster and more widely used around campus. I had asked a few of the team members if there would be any problems. Andy had previously suggested it and thought it was a great idea. Dave saw no problems and was all for it. Nobody else had any issues.

Except Chris.

I didn't ask Chris. Not because I didn't want his input, but just because he wasn't around when I was asking people. I had a majority of the team, including the chief 3D design guy, Dave, saying it was not a problem.

After I sent the e-mail out alerting everyone of the change, I got a response from Chris.

"Goddamnit, you couldn't wait until at least the car was built?"

Chris frequently made issues out of irrelevant things, but he was usually more passive aggressive about it. I called Bobby to see if maybe Chris was joking. He was not.

I sent him a reply.

"I talked with several of the team members and nobody saw an issue with it. Aside from that, Dr. Sommer wanted the software upgraded, and it's his lab, so your opinion on the matter means very little. This is a non-issue."

I went to sleep, somehow expecting that was the end of it.

The next morning I had a reply from Chris.

"Well good for you, IT guy. I wish you were as passionate about your duties as chief engineer."

What the hell is his problem? I thought. I was meeting all of my duties. I was even plowing through his condescending bullshit to oversee the design and manufacturing of his system. I spent every meeting listening to his sarcastic jabs at my expense every time I asked him about his progress.

I decided to take a shower and cool down before driving to the shop. After the shower, I was still mad. After the drive to the shop, I was even more upset. I walked into the shop and shut off the welder that Chris was using.

"Well? What is your problem with me?"

"I want you to do your job. You're not doing anything as Chief Engineer, you've got all these little side projects you're always working on —"

"What exactly is it that I'm not doing? I'm not keeping things on schedule? Every time we have a meeting you show up 15 minutes late, complain about the meeting, and respond to all my questions with some smart-ass response. I can't keep a schedule because you don't respect my attempts to keep a schedule. You sabotage all my attempts at completing my duties and now you want to complain that I'm not doing them? What's the real issue? I responded to your problem with the software and you just shot back another complaint," I responded, arms waving and fingers pointing.

"I was half joking about the software thing," he said quietly.

"Well, it's obviously not the software, and probably not the other thing, so what's the real issue?"

"You're not doing your job, you didn't send in that rules clarification. You don't know how much you hurt team morale when you refused to send that in," He responded calmly.

"The fucking rules clarification? That was a month ago, and besides it's not my responsibility to send it a clarification about some fantasy rule that you made up."

"It's your responsibility to oversee the systems."

"Yes, oversee your system, if you overlook a rule it's my job to catch it, but it's not my responsibility to verify that *your* design meets the rules. It's *your* design and *your* responsibility to make sure it meets the rules. That's part of being the designer."

"Well, I guess we'll just keep doing what we're doing and get stuff done," He reached back and turned on the welder.

I went upstairs to cool off. I thought about what he said. I thought for a long time. There have been times in the past where another team member and I had an argument, but it was almost always some fault of mine. I liked to think that I had grown a bit over the years. I liked to think I was better at dealing with people. Where was my fault in this? In retrospect, I should have just sent in the rule clarification, but that shouldn't have led to a month worth of indignant contempt.

Bobby came into the computer lab.

"What is Chris's real problem?" I asked.

"What do you mean?"

"Why is he being an asshole to me? He said it was the rule clarification, but that was a month ago."

"I don't know. He was a little mad about the software upgrade."

"Yea, he said he was half-joking about that, whatever that means. Has anyone else said anything about the rule clarification?" I asked.

"I haven't heard anything about it. I think he's just in one of his moods. Like you said, the rule clarification thing was a month ago."

"Well, I am completely sick of dealing with his BS. I'm tired of taking crap from him just because he's in a bad mood every other day," I said, cementing the thought in my mind.

"But you have to be able to avoid having a conflict that affects the team—"

"That's what I'm afraid of," I cut off Bobby, "I'm afraid he's going to screw the team over just to take a jab at me, or someone else. He's already done it once with that Sommer thing."

"I know, and I told him after that whole thing that if he did anything like that again, he would be off the team." There was a brief silence. "If you have a problem with Chris, let me know and I'll try to mediate. But you need to—"

"I have a problem with Chris," I interrupted. "Mediate or I'll…" I stopped, not wanting to make angry threats. Bobby had a look like he didn't have time to deal with my shit anymore than I had time to deal with Chris's. I shook my head, turned around and walked out.

If Chris had left the team right then, I would have been happy. In fact, I was seriously considering trying to get him to quit the team by relentlessly ridiculing him in front of the team every time he made a smartass response. It wasn't until four days later that I abandoned my idea of getting him to quit the team. He sent me an e-mail with his design report attached and added this at the end:

"On a more personal note, I am sorry for blowing up the other day. I have been under a lot of stress with school and other things, and it was unfair of me to lash out at you. I have considered you to be a friend and hope we can be so again in the future."

I had cooled down enough to not be mad at him, but I wasn't about to just move on this time. I replied:

"I sincerely apologize for yelling at you, and for refusing the request to send in the rule clarification. I would very much like for us to be friends and teammates, but it will require more than just an apology this time. If you want some very useful information, your Achilles' heel(s) is that you are quick to attack other people's skills and character (frequently when they are not around), and yet your feelings get hurt far too easily. The former is far worse than the latter."

The next Tuesday night meeting went smoothly. I asked Andy about the uprights and Chris chimed in.

"What are you going to do about pressing in the bearings?"

Andy responded, "I was just going to press them in."

"No bearing sleeve?" Chris replied indignantly, "You know, you could just beat it in with a mallet, it'll be fine," he said sarcastically.

"Don't be a smart ass, Chris," I said.

He stopped talking. While Andy and I discussed his deadlines, Chris got up and left the meeting.

We finished the meeting and I started to head downstairs to machine the foam for the wings.

"Not yet. You, me, and Chris are going to have a talk," Bobby said, "There has been some tension between you two and we're going to figure it out right now."

For the next 15 minutes, I bit my lip and listened to Chris tell me why he was so upset with me.

"You always read something about the engine system or something, and then you quote it like and expert..."

"You have a double standard for me, you called me out for being a smartass, and then you didn't say anything to Kyle and Ricky for their little side conversation..."

"You aren't doing your job and you hurt team morale when you didn't send it that rules clarification."

Do I quote things like an expert when I have just learned the information? Sometimes, maybe, but that was at worst an annoying personality trait. Did I treat Chris unfairly? Did that rules clarification really hurt the team morale? I asked Kyle and Ricky, separately, about all these problems and they both came to the same conclusion.

"Yea, we were a little pissed about the rules clarification, but Chris just gets furiously mad about everything, stuff that most people would forget about after five minutes."

Bobby had made us sit down and talk out our problems in hopes that we would work them out and get along, but that's not what happened. I left the meeting more defensive and angry. I took things personally; when other team members said things about me, even jokingly, I always did some personal reflection. But Chris was always mad at me. He was almost always in a bad mood when I was around and rarely responded or communicated with me without some sort of resentment. I put up a wall.

Fuck him, I thought, *and fuck everyone else; we've got a car to build and we don't have time for this shit.*

"Since Sommer's been gone, you've been taking the brunt of it," Kyle said, "He didn't used to be like this; he used to be happy all the time. But something happened and now he just gets irate at the littlest thing."

"Well I'm done," I said, "I'm done being his friend and I'm done caring what he thinks. I spend too much time on the team to put up with his shit." I looked at the clock on the wall. "And now I'm going to be up all fucking night machining the wing molds because Bobby wanted to have a marriage counseling session."

Kyle sighed, "I hate losing friends."

"I don't think I'll miss this one."

"Well, if it doesn't kill you, it can only make you stronger, right?" Kyle said, trying to make it a positive situation.

"Life's school of war." I said quietly to myself.

"Huh?"

"Nothing, never mind." I said, staring pensively out the window.

"You should come out with us to the club tonight, get away from the car."

"Huh, right. No, I'm not really the dancing type. Or the socializing type. In fact, everything about that sounds... the opposite of me."

"Fun?" Kyle said with a smile.

"I was thinking 'extroverted'. Anyway, I'm going to go start on the wings."

I went downstairs and began preparing the CNC mill. It was going to be a long night. I was tired. Tired of all the bullshit, tired of working eighty hours a week on a car just to get bitched out for nothing. Tired of being completely broke, tired of living in a shithole. The only friends I had were the people on the race team who I was tired of seeing and who were tired of seeing me every waking hour of every day. All my old friends stopped calling; I blew them off every time I talked to them because I had to meet some deadline for the racecar. And for what? What was the point? So I could get a job that pays $45k a year designing seatbelts for the new Civic? I wasn't even sure what I wanted to do anymore.

Bobby was wearing on my last nerve, siding with Chris on all his bullshit. He was barely passing classes, but I wasn't even sure anymore if that was an indication of the amount of time he spent on the team. The rest of the team seemed like they were breezing through the year. Kyle had a Lexus, a girlfriend and lived in a huge house with a hot tub. He somehow managed a 3.8 GPA and was going to graduate school in England for Racecar Engineering in the fall. My GPA was barely above getting kicked out of the engineering college, and well below the minimum required to get a job designing seatbelts for Honda. I had nothing to fall back on. All my cards were on the table, and they all said FSAE.

I looked at the clock on the CNC computer. 10:15 and eight hours of machining to go. Ricky and Kyle would probably get to sleep the same time as me, except they would be winding up a night of drinking and dancing and sitting in the hot tub.

"Fuck it," I said quietly to myself. I bumped the emergency shutdown button on the CNC with my fist, shut off the computer, turned off the lights and locked the door behind me. There has to be more to life than deadlines.

8

Soccer Four Square

"I don't know driving in another way which isn't risky. Each one has to improve himself. Each driver has its limit. My limit is a little bit further than other's." – Ayrton Senna

I pulled into Kyle's driveway hoping they hadn't left yet. His car was still there, as well as a few others. I walked up to the front door wearing my hastily assembled outfit consisting of my only button up shirt, and one of the last stain-free pairs of jeans I had left. I didn't have nice clothes, just clothes that hadn't gotten dirty yet.

I rang the doorbell and Kyle opened it with a look of surprised confusion.

"Matt?" he said.

"Is Matt coming tonight?" Ricky said loudly from the back of the living room.

"I assumed your invitation was still open."

"Hell yea it is, come on in. We're finishing up our pre-game."

I learned that pre-gaming is getting half drunk at home where it's cheaper, and then finishing up at the club. Ricky offered to make me a shot, but I insisted that I was a beer guy. Kyle introduced me to the other people that would be going with us; their roommate and a couple other people. I didn't like social situations or meeting new people, so I took up Ricky's second shot offer to take the proverbial edge off.

We packed into two different cars and headed for the club. When we arrived, I was still feeling tense. Ricky suggested I try a Red Bull Vodka, but I was happy with my beer. Still, I didn't want to be the stuffy boring guy, so I joined in their round of shots at the bar. I had a few more beers, talked with some of the other people we went with, and had a good time drinking and not being in the shop. Still trying to relax and fit into the group, I moved to the recommended Red Bull Vodka. Everyone else had been dancing since we got there, and they had all tried to get me on the dance floor. Going out to a club was a stretch for me, but dancing was something I was not going to do. There were many things that I was not, and chief among them was a dancer.

Seven Red Bull Vodkas later I was a dancing machine. Kyle was a really good dancer, which was kind of intimidating. Fortunately, for all his confidence and bravado, Ricky was not. A quick scan of the room revealed a sea of other bad dancers. I didn't really care anyway, I would never see these people again, and Kyle and Ricky were well informed about my lack of dancing abilities. So I danced, and I drank, and we laughed and joked and drank some more. By the time we got back to the house, I had forgotten we even had a racecar. By the time we finished the bottle of Vodka, I had forgotten I had a name. I crashed on their couch for a few short hours before making the painful drive home into the morning sunrise.

I woke up at 10:40, late for Cherokee. I decided to skip class; better never than late. I went to the sink to brush my teeth and stood staring into the hazy car side view mirror that I had fastened to my bathroom shelf.

For the past several weeks, the race team had seemed like a

battle. A battle between Chris, and school, and deadlines, and any sort of social interaction outside the shop. It was a series of battles; a war that I had fabricated using only my innate ability to funnel all of life's possibilities into a racecar competition and a zero dollar bank account balance.

I thought about what Kyle had said the previous night. "If it doesn't kill you, it can only make you stronger." I appreciated what he was trying to do, but I hated that quote. It's not that the quote lacks any inherent quality, just its popular use. People throw it around like an affirmation for when life takes the opportunity to slap you in the face. It doesn't really mean anything, and it's certainly not true the way people use it; strength does not automatically show up in a neat package on your doorstep as a consolation of hardships. Lessons have to be learned, and personal growth and all that shit. The whole quote is "Out of life's school of war: What does not destroy me makes me stronger." It seemed appropriate, though; if anything was a school of war, FSAE certainly was. The point was to learn something, and it existed in the context of competition. We were interested in learning, but really we were there to win. Learning was ancillary, incidental, a bonus and a good excuse for us to ask the university for money. But it was there, we learned a lot, and our drive to win required us to become stronger in the face of setbacks. Out of FSAE's school of war, what doesn't kill me only gives me the opportunity to make myself stronger.

In any case, I knew what my problem was and I didn't need Nietzsche to tell me. I had chosen every situation I was in. I had chosen to move into a shithole apartment, to blow all my extra money on my Dodge Dart, and to spend all my free time building racecars. I had no friends left outside of the team because I blew them off 90% of the time and took the last 10% for granted. I had taken the responsibility to make those decisions and I had to take the responsibility to live with them. Kyle was in the situation he was in because he had made different choices and made other sacrifices. He worked his ass off all day, every day and I knew it. He was a great friend and I should hope for the best for him. Bobby was also dedicated to the team, as much as anyone. The team was often a frustrating situation, and I had started to resent the other team members because of my own stress and my own problems. Something needed to change, and it wasn't them.

As for Chris, well Chris was just an asshole sometimes. But the truth is that I was an asshole, too. Where Trey or Kyle or Bobby would have shrugged it off, I let my ego escalate small arguments into battles. When Chris tried to defuse our arguments, I took an arrogant jab at him. It was hard to step away from the situation and look at what I did wrong, but the emails were there, preserving my arrogance and unashamedly showing me that I said what I shouldn't have said. In a lot of ways, Chris was defending the team and our progress and I was defending my ego. He did his part as well, but it takes at least two people to start a war.

We wouldn't be best friends; in fact, it was several weeks before we talked at all. We were teammates, and we made good teammates. He did his job and he did it well. He was self-sufficient and would get his job done with our without my help. We came around slowly, carefully talking about the car and avoiding the landmines we knew were there. It wasn't worth the arguing and the fighting. It would only make the car slower and our lives miserable.

We continued manufacturing the car through the winter break. I had stopped by the computer lab on Christmas Eve to pick up something on my way to my parents' house, and Dave was there, quietly working by himself. I expect he had been there most of the day; he liked a computer lab with no people.

At the beginning of the spring semester, my student loans showed up and I was back in the black. I added a signature student loan and carefully budgeted my money. I planned to have it last through the semester and two months into the summer in the likely event that I would still be looking for a job after graduation. I even budgeted for the last few things I needed to get my car running.

Things on the team started to fall into place as well. We had successfully signed up for the California competition with none of the drama we had enjoyed while trying to sign up for Detroit. Registration opened for both competitions at the same time, but teams could only register for one of them for one month. Then, after a month, if spots were still open, you could register for both. They did this to make sure everyone got to go to at least one competition. Detroit had filled up in a matter of minutes, but California still had eight open spots. When the day came, we had a network of registers ready to go. Ricky and Kyle were at home with their credit cards. I was at my apartment with my roommate's credit card, and Bobby was at school using the engineering department's card. Whoever got us registered first would be paid back immediately. We were all going to try, at the same time, on different computers. Ricky got us registered in less than a minute, before the competition was filled three minutes later. We were going to two competitions. It was nice to know that all the pressure wasn't riding on only one.

I hadn't really thought about firing Phil since I had originally suggested it, but it was clear that it was not the right thing to do. He had shown up to the meetings and done his job; he was on schedule with his parts and the cost report was done well and submitted on time.

The day the trailer was towed, it should have been clear to me that Phil was a big benefit to the team. He was Mr. Miscellaneous; He did all sorts of stuff like getting the trailer out of impound. Stuff I didn't want to do and didn't have time to do. There were a few of us, Bobby, Kyle, Ricky, Chris, Dave and I, who were more or less obsessed with the team and the car. So it became hard to accept that some people weren't obsessed. Even when someone was working 10 hours a week on the team, it was

95

easy to think they were doing nothing, but that 10 hours was a valuable contribution, and about 10 hours more than they were required or paid to do. Phil was a 10 hour guy, and he needed a 10 hour job. The best thing about Phil was his attitude and, for that reason alone, he was worth having around. He was always in a good mood, always positive, and I'm sure he would have taken a bullet for anyone on that team. Anyone at all, probably, random strangers even. He was just that kind of guy. Also, I had realized that it was team management's responsibility to make sure he was on schedule. With Chris and Dave being defensive about my deadlines, I had backed off bugging the team members about their parts and systems. I should have kept up the schedule, but approached the team members individually and with a balanced concern for their own schedules and the team's schedule. The team had agreed to place Phil in charge of the cost report, and Bobby had done a great job of making sure the cost report was on schedule and up to par. Bobby was a better manager than I was, no doubt.

Into the spring semester, as the racecar started to take shape, the team seemed to take a new shape as well. We were all having more fun, hanging out and being less worried about the car all the time. Kyle, Ricky and I went out to the club a few more times and I made new friends with their friends. I even danced relatively sober a few times. Don't tell anyone. As part of my anti-anti-social effort, I started going over to Kyle's parents' house with a few of the other team members to watch the last few OU football games and the Super Bowl.

My personal stress level had gone down since I stopped incessantly nagging and worrying about the schedule and started spending a small bit of time away from the car. Overall, my productivity went up. Working 60 hours a week on the team is usually more productive than working 80; mistakes get made, and you start to spend hours at a time staring at the screen, grinding your teeth, and trying to remember why you haven't quit yet.

The truth is, no matter how hard you work, if you're not having fun you're probably not going to win.

On occasion, when there were several of us at the shop, we would get together and play soccer four-square. This was our version of four-square with soccer rules. In other words, no hands. It was fun, and it got everyone to relax. More than once we had to fashion a retrieval device to get the soccer ball off the roof, which was an engineering challenge itself, considering the 20 foot roof.

One night, after he missed the ball and lost the king square, Wes grabbed our ball, yelled, "This is Sparta!" and kicked it as hard as he could. None of us knew what just happened; we were all just shaking our heads and laughing.

"What the hell was that?" Bobby asked, laughing hysterically.

Wes just shrugged his shoulders.

"Alright," I said, "let's get some work done on the car. Wes, go find our soccer ball."

We went back to our respective projects. Ricky was working on a go-kart that he had brought in. It was his cousin's or something, a half rusted yellow thing with a tear in the seat and a flat front right tire. Ricky was being the good family member and fixing it. Everyone was okay with him working on it because the car was on schedule and Ricky had all of his stuff done. Also, I had taken my car up to the shop more than once to help get it in running condition.

"Has anyone seen Josh up here," I asked, "he needs to get the electronics done like a week ago."

Wes answered, "He shows up to work on it for fifteen minutes at a time and only once every several days. He's not making any real progress."

"He needs to get here and get it done. It's his responsibility," Kyle said with a hint of frustration.

Ricky, who had done the electronics the previous year, said, "He needs to fix this," pointing to a wire that was running towards the pedals. "These spade connecters are crap, and the wire doesn't have any slack. It's a terrible design."

The wire he was talking about was for the brake over travel switch, which shuts off the car if the brakes lose pressure. It was a safety requirement and something Josh had done as soon as the pedals were in the car.

"But the electronics system is a thankless job," Ricky continued, "you can't really start on it until the car is mostly done, and then everyone expects you to have it done in a day. Why do you think I didn't do it this year?"

Ricky was right, but it was clear that Josh wasn't going to meet his deadlines. He was a 10 hour a week guy like Phil and the electronics system was too much for someone spending only 10 hours a week. Lately, he had been spending less than 10 hours on the team, and spending more time at his job. He was enthusiastic, and when we offered him the electronics system lead he took it, but we should have waited to see if he would have put in the effort himself. We had made the same mistake with Ralph, who had left the team long ago, but I expected more out of Josh because he reminded me of me. But he wasn't me, and I should have stood back and let him beat his own path. If he really wanted it, he would have made it to electronics team lead eventually, and he would have deserved it and given it the time that it needed. If he didn't really want it, we would have known then, when it didn't matter so much.

In any case, most of the rest of the team was on schedule, so we could all pitch in to help get the electronics done.

But that was for another night. My present task was to finish the supports for the rear wing. The supports were made with carbon fiber tubes bonded to aluminum fasteners. I prepared the surfaces, mixed the adhesive, and glued the pieces together in their appropriate places. While I was waiting for the adhesive to set, I noticed a small fire in the parking lot with Wes hunched over it. I walked over to investigate.

"What are you doing?" I asked.

"I just wanted to see if this gas was still flammable. It was in the go-kart for a long time. It doesn't even really smell like gas anymore," he said, handing me the gasoline filled cup.

"You remember back in middle school when people used to play with fire," I said, before smelling the cup, "and there was always that one kid that would take it way too far and almost catch the neighborhood on fire? Yea, I was that kid."

For some reason, Wes's response was, "You know there's a whole bucket of gas left in there." He pointed towards the go-kart.

Five minutes later the entire parking lot was on fire. I had just lit the whole bucket, which promptly melted letting flaming gasoline pour out in all directions. I grabbed a cardboard box to smother the flames, but the box just caught on fire. Then, stomping on the cardboard box, my shoe caught fire. Wes picked up a fire extinguisher and took care of it. Before the fire was even completely out he yelled, "What was that! What the hell were you thinking?"

"Wes I just told you... what did I tell you, what did I just tell you?" I couldn't stop laughing. I knew the fire would be fine, the parking lot was huge and there wasn't that much gasoline. Besides, I had a lot of experience with these kinds of things in middle school. Wes kept shaking his head, fending off the urge to laugh, and keeping an angry look on his face for another minute before giving up.

"You ass." He said with a smile.

I cleaned up the mess and got back to the project at hand. While I was busy setting the parking lot on fire, the adhesive on my wing supports had cured to the okay-to-touch point. I picked up one of them and set it on the car where it would later be bolted.

"What do you think?" I asked everyone.

"Are those the carbon fiber ones?" Kyle asked enthusiastically.

"Yep."

"They look kind of flimsy," Wes said.

"Yea, that's carbon fiber. I put test pieces in the material test machine and they don't buckle 'til 400 pounds."

I purposefully didn't pass them around, because when engineers are confronted with a very lightweight and fancy looking thing, for some reason it is a natural reaction to twist it and bend it to see how strong it really is. While these parts were perfectly fine for use on the car, they were not designed to be twisted and bent by curious engineering students.

"Did you just glue those aluminum pieces on the end?" Brett asked.

"I don't like to think of it as glued so much as bonded," I emphasized bonded with finger quotes.

"Well if it has a fancy name, then we should be good," Wes said with a smile.

I scowled at Wes for a moment before coming up with my wittiest response ever, "Your mom has a fancy name."

"Are you calling it good?" Ricky asked.

"I'm calling it good," I said confidently, "It's actually a structural adhesive, it's got a shear strength of... some huge number, I forget, but it's almost greater than the bolts holding the wing supports to the frame. They're not going to break. They are, however, very fragile in bending and very notch sensitive, so when they're not on the car we are all going to treat them like egg shells, right?"

Bobby gave the response I knew was coming, "You have steel backups, right?"

"Of course, I'm not stupid." I put the wing supports carefully into a box and placed them in the back corner of the shop, away from twisting, bending hands. I looked at the clock on my cell phone and decided it was time to leave. "Well I think it's about time to head out."

"You need to get up early for Tennis class, Matt?" Kyle asked with a smile.

"Ha. No, I stopped going to tennis a month ago. I decided my easy semester of easiness needed to be a little easier."

"Really, are you going to fail it?"

"Yea, but it doesn't matter, it's a pass/fail class so it doesn't affect my GPA. All it does is put me on probation for financial aid next semester, and since I'll be graduating this semester, it seems like that won't be a problem. I would like to get some sleep, though."

We all agreed and called it a night.

9

Hawaii Local Time

"The price of winning is always the reduction, if not the elimination, of play time. However, since racing is basically playing any way you want to look at it (real people make their livings by doing something they hate), we can't bitch too much."
– Carroll Smith

During one of my seldom trips home, a high school friend asked me why I had chosen, among all of life's opportunities, to spend all my time on a racecar. I didn't have an answer for him. I had never thought about it. Amidst all the chaos and stress and frustration, during all the sleepless nights and last minute deadlines, I never asked myself why I was doing it. That was my gift, from me to me. Most college students switch majors three times, and spend years wondering what they want to do before spending the last few months wondering what they're going to do. I was already doing it. Still, it got me thinking. Why the team at all, and more importantly, why dozens of hours every week. I wanted a job with a car company and the team would look good on my résumé, that's why I showed up for the first five hours a week. I showed up for the next five because I loved racing. I spent all the rest of the time because I wanted to drive, I wanted to be the team technical leader, and I wanted to win. I wanted to be the guy in the car, the guy in charge of the car on the team with the best car.

The allure of that position had lost most of its luster by mid-year. Somewhere between our faculty advisor vacancy, my near failing grades, and fuming arguments with team members, I knew that a big trophy wouldn't make it all worth it. Leaving the team midyear was out of the question; I would have stayed until the end of the year no matter what. But by spring, I had found new enthusiasm, new motivation to show up more than ten hours a week. I was actually having fun on the team, playing soccer four square late at night and having a social life. Granted, it was with all the people on the team, but it was much better to have a group of friends who also worked together than it was to have a bunch of pissed-off team members laboring at the behest of the unsympathetic Master Schedule. I was trying to be a better team member and a better person; a friend more than a manager, or at least a friendly manager. My answers to the why question were all the same as before, but some of the answers had shifted importance.

Unfortunately, "why" I wanted to engineer racecars was not the same as "where" I was going to do it, or more importantly "who" was going to pay me to do it. Midway through the spring semester, most of the graduating team members knew what they were going to do after college. Chris was staying for graduate school. Kyle was going to be pursuing a graduate degree in racecar engineering in England. Andy was getting a job in Texas with an electronics company and he would be leaving immediately after graduation, not even staying for the competitions. I, however, was still looking for a job. Over the course of a month, I had sent my résumé to every major racing team in every major racing series in the United States, two auto manufacturers, and three racecar part suppliers. I had sent it to teams I thought I had a good chance with, as well as the teams I had very little chance with. I had very few responses and no positive ones. I didn't want to go to graduate school, which was convenient considering my GPA was well below the requisite for any school that didn't have "cosmetology" in the name.

On the positive side of things, the race team was no longer without faculty representation. Dr. Siddique was our new faculty advisor. I was a little worried at first that it would be a difficult transition to a new advisor partway through the year, but he recognized that we had everything relatively under control and he didn't insist on anything. He signed the papers that needed to be signed, came to our meetings, and helped us when he saw an opportunity.

Meanwhile, we continued to work hours upon hours to get the car done on time. Bobby and I filled in our car information for the event programs that would be handed out at the competitions. Of course, instead of entering the actual information, we thought we would be funny.

"How about 'Rich Mahogany?'" Bobby said, typing it into the 'bodywork' text box.

"For 'Wheelbase', type 'At the Wheel-Lair'," I said, laughing.

Chris and some of the other team members came into the computer lab and made suggestions. By the time we were done, our list of car information had no actual car information and included:

BRAKES: The Midas touch
DRIVE: Hope and Inspiration
ENGINE: Yes
OVERALL LENGTH: Ask your mom
SHIFTER: Minty-fresh
TIRES: Round
WEIGHT: We weight for no one

For most college students, spring break is a time to relax and take a break from all the hard work. For us, it was a chance to spend all our time on the car instead of just most of it. We logged a lot of hours on the car during that week every year. I am sure that I had spent over a hundred hours working on the car the previous spring break. It seemed like we always set our running car deadline at the end of spring break, leaving us to frantically build the car all day and all night for a solid week while our peers were partying MTV style in Cancun. Setting it a week before wouldn't have helped our already distressed grades; that's when midterms were, and a week after wouldn't have given us enough time to do all our testing. Optimistically, perhaps, we had set our deadline for two weeks before spring break for the '07 car. We didn't set an actual date, but as the week before spring break approached we settled on Sunday. Specifically, Sunday night at midnight. This would leave us with a week to actually study for midterms and a week after to party MTV style or, more likely, work on the car without the deadline stress. Maybe only 60 or 70 hours that week.

So on the deadline Sunday, we gathered at the shop at 11am and started down our list of what was left to do.

"Stub shafts. Is Chris working on the stub shafts?" I asked Bobby.

"Yea, he's using the CNC mill at the machine shop," He replied.

"Excellent," I said. "Brake system?" I looked at Kyle.

"Huh? Oh, I need to run the hard lines and bleed the system," Kyle said.

"Is that gonna be done today?" I asked.

"Maybe."

"...okay," I said, expecting more of an answer.

"If I have the car by myself, it'll be done, but I can't do it if everyone is hovering around," Kyle added with a hint of defensiveness.

"Fair enough," I continued with the list, "Cooling lines. Doesn't Dave usually do the cooling lines? Where's Dave?"

Everyone was silent.

"I haven't seen Dave since... a few weeks ago," Bobby said.

"I haven't either. Where the hell is Dave?" I asked louder, "Has anyone seen him lately?"

"I haven't seen him in at least a month," Trey said.

He was so good at blending into the background that we hadn't even noticed he was gone.

"Yea, I think he quit on us," Ricky said understatedly.

"Awesome," I said, "we don't have Dave anymore."

I took a deep breath and thought about our schedule, and about how I had long ago expected that we would have to finish the year without Dave.

"Fuck it; we'll do it without him," I said decisively, and then went to the next thing on the list, "Ricky, can you help Josh finish the electronics today?"

I finished going through the list, making sure everyone had something to do. As I was about to wrap it up, Kyle asked me, "Are the wings going to be finished?"

"Uhhhh…maybe. I don't really count that as part of the running car. I've still got to finish the fuel system which is, decidedly, part of the running car." I wouldn't have the wings done, and I knew it. I had done a pretty good job, I thought, to get them as far as they were. Especially considering that I had basically picked up the fuel system. However, the team wasn't concerned with excuses and 'pretty-good'. I had expected all of them to get their stuff done, and they had every right to expect me to get my stuff done. The wings were just not going to happen this weekend. They all probably knew it and were all fine with that, as far as I could tell. "Okay, T minus…" I looked at the clock and did the math in my head, "twelve hours and forty six minutes. Let's get it done."

We worked all day, finishing up all the things we had on the list and adding to it as we remembered details we had forgotten about. We always forgot some details, no matter how many cars we built. It was the small things that killed us in the end: the nuts and bolts, the hose fittings that just never crossed our minds.

We took a dinner break and played a bit of soccer four square to relax before getting back to our impending deadline. Ricky and Chris had helped weld together our fuel tank, and I had as-sembled the system for a leak test using water. We liked to find the leaks with water instead of gasoline so later when we welded the holes closed the tank didn't explode.

We had some foresight.

Unfortunately, the tank had about five leaks. I circled the leaks with a marker, drained the water, and asked Ricky to weld up the holes. We repeated this two more times with the fuel tank before it was fully sealed. Then we did the same with the radia-tors, which we found to be leaky upon installation. I kept my eye on the clock, and as it approached midnight I started to realize that we weren't going to make it. That's the problem with finding forgotten details: they usually lead to other forgotten details. At 11:55 I took the clock off the wall and set it back one hour. As I put it back on the wall, I said, "Alright, we are now on Mountain Standard Time and…" everyone laughed "…we have one hour and five minutes left."

We spent the next hour trying to finish up the wiring.

"Okay, try it now," Ricky said, looking down at his multi-meter. Josh reached up and pushed the arm button.

"Smoke, smoke, SMOKE!" Wes frantically pointed to the electrical box. Josh quickly shut the car off.

"That's fine," I said calmly, "every car needs at least one elec-trical fire."

We patched up the wiring and continued our mission.

Phil, setting the clock back on the wall, and with a big grin, said, "We are now on Pacific Standard Time."

Just under three hours later, a full 15 minutes before our deadline in Hawaii local time, the car was ready to drive.

"Bobby," Chris said, "get in the car, drive it around the park-ing lot and let's go home."

Bobby got in the car, put his helmet on, and started it up. The fuel map wasn't tuned correctly, so it wouldn't rev up all the way, which was just as well because the rear differential brace was still missing, making hard acceleration unadvisable. Bobby carefully drove it around the parking lot, and we all gave a tired but heartfelt 4:55am cheer. It didn't run great, our clocks were five hours off, and the shop looked like an airplane had crashed into a soccer riot, but we had done it. We went home to prepare for our week of midterms, which began about four hours later. Once again, studying lost out to sleep, just as sleep had lost out to the racecar. You have to have priorities.

A couple of weeks after our first drive, we were still testing and tuning and didn't feel quite good enough to put a bunch of miles on our new car. All the same, the time had come for us to pick drivers for the different events, so we took out the '06 car instead. We wanted to get driver selection taken care of because driver training was extremely important, and since we would be doing full car tests in the near future we wanted to test with the drivers that would be in the car for whatever event was most closely related to that test. We set up the different driving events and everyone that had put a lot of time into the car got a chance to try out. The fastest people got the spot. We had an idea of who would probably drive the events, but it was good to let everyone try. For one, it was good for team morale, but also we might have been surprised by someone's speed and there was only one way to find out. We had everyone run the skidpad and acceleration events twice, just like they would at the competition. Our drivers were Ricky and Chris for skidpad, and Bobby and Ricky for acceleration. The other two events, autocross and endurance, were basically the same except that autocross was one lap, and endurance was ten laps per person for two people. So we had everyone drive ten laps each, with the first lap counting as the autocross lap and the total of all ten for the endurance. Chris had almost the exact same time as me for the ten laps, but he hit a cone which added two seconds and put his time behind mine. Kyle and I would drive both the autocross and endurance events.

10

The Final Touches

"High achievement always takes place in the framework of high expectation." – Charles Kettering

The 2007 car was the best looking car we had ever made. We spent some time thinking and arguing about the colors, but most of the car was purpose built to go fast around a racetrack with aesthetics a secondary criterion, at best. Major racing teams have millions of dollars and millions of hours going into the car designs, so that every single component of the car is considered for its ability to help make that car go around the track faster. Even the paint jobs are specifically and carefully designed to sell the product advertised on the car. Yet, with all the effort put into engineering a machine to perform a specific task, it ends up looking like a piece of art. Art is often considered to be an object that has no function and only aesthetic value. I believe that if you go to the extreme of function, fully opposite any attempt at aesthetics for the sake of aesthetics, you end up with something that is deeply and inherently beautiful, transcending aesthetic value to a beauty that exists not only because it is functional, but also only because it is functional.

We got a little closer to that ideal this year. The wings helped tremendously, and the level of detail that the team members brought helped to add purpose of design to every part of the car. We actually got a few weeks to enjoy it, too, before tearing up the car at the competition and in transit to and from. Historically, our team was trying to finish the car well past the running car deadline. Usually through finals week, then all night before leaving for Detroit, then a few added touches in a parking lot on the way to the competition. There were always some teams finishing their cars in the paddock during the first two days of the competition, as was the case with our team several years back. One of the most important things we learned in those years was that a great car that doesn't run is much worse than a good car that has a lot of testing.

During testing, we got a lot of help from Kyle's parents who let us use some of their racing test equipment. Our budget was limited and especially strained as we approached the end of the year, so borrowing equipment was a huge help.

Testing is extremely important in auto racing. The car never works just the way you expect it to, or just the way the computer says it will. Backing up simulation with physical testing is to engineering what staying awake is to driving. When you go into the bathroom at an engineering office, there is a sign on the wall that says "Employees must wash their hands before correlating computer simulation with testing."

A few of the components on our car were hastily made in 'good enough' condition to meet a deadline, just to be remade later when there was spare time. Parts of the suspension were made from steel to meet the rolling chassis deadline, only to be remade from titanium later. It seems like a bunch of extra work, but time driving the car is extremely valuable and well worth the extra manufacturing time.

This year, we even had enough time to go overboard with our push-bar. The push-bar, a requirement for all teams, is a structure that attaches to the car so that it can be pushed from behind. A couple years before, the rules committee made it mandatory with a couple functional requirements: it had to be usable (and used) by two people and it had to be able to pull the car as well as push it. We thought it needed some extra complexity. It started out as a suggestion that we add a bell from a bicycle to let people know we were coming. While we were picking up the bell, Wes and I decided it needed a few extra parts from the bicycle section, so we added some purple handlebar grips complete with glittery pink and purple tassels and a pink basket. Later, Kyle, Ricky, and I went to a hobby store to pick up some fake flowers for the basket, a teddy bear, and a purple toy windmill. It was spectacularly ridiculous, truly a triumph of engineering.

Finals week was a breeze for me that last semester. For four years, we had been trying to build the car at the last minute, cramming for finals at the last second, and then going back to the car for another round of sleepless building. A couple years before, Chris had stayed up all night working on the car, went to one of his final exams, and deliriously spent twenty minutes answering questions about Thomas Jefferson before he realized that he wasn't taking American History that semester. He had arrived in the testing room two hours early.

Our last official meeting before the competition was the Tuesday of finals week. We were far enough along on the car that everyone's finals week would be more laid back than usual, at least on the racecar side. We had told everyone that the meeting was optional and that they should focus on their exams. Still, most of the team showed up.

"So I got an email from someone in charge of putting together the event programs," I said to the team, standing behind the podium, "Some people there were not amused by the information we added. Apparently, they don't believe we have rich mahogany bodywork, or a minty-fresh shifter."

"Are they not going to print it?" Bobby asked.

"Not only are they not going to print it, they are going to send a copy of it to our faculty advisor," I answered, "I don't think Siddique will care, it's not like it was offensive. Honestly I don't know why they were so upset by it."

"I think it was probably the 'Ask your mom' answer we put to 'Overall Length'," Ricky suggested.

Kyle added, "Yea, we should have probably left that one out."

I shrugged my shoulders and moved on to the next subject, "So as you all know, I've been working on the poster boards for the design event. The poster boards, for those of you who don't know, are just an easy way for us to display important graphs and charts to the judges in the design event. It's so you don't have to be flipping through a notebook or a bunch of files on your computer when you want to show test data. It's also good to have the data for stuff you know they are going to ask about." I reached behind the podium and picked up the aerodynamics poster board I had made. I decided on the middle school science fair approach, which was to buy foam board and cut and paste graphs and charts with construction paper frames behind them. "So this is what I've made for the aero poster. It's not a full blown printed poster with a glossy finish like some teams, but as a lot of you know, some of the teams that do the best in design, like Cornell and Penn State, take this approach. So basically, I know everyone is busy with finals, so wait until you're done with your tests and then make up some graphs and charts for me and we'll do this... Elmer's glue and safety scissors thing. Everyone cool with that?"

"That actually looks really good," Kyle said.

"I know," I replied, "which is weird considering that I have the artistic talent of a drunken third grader."

"Andy is not going to the competition with us, because he got a job," I said with satirical disgust, "So Andy, if you wouldn't mind getting together a few things for the design boards before you leave. Oh, and go over them with Kyle and Trey, they're going to be taking care of suspension in the design event."

"Is there anything in particular you're looking for?" Andy asked.

"Not really, just remember what the judges want to see." I tried to think about specifics for Andy, but I didn't know enough about the suspension. "It's your system and you've been to the design event before, so make it count. Computer analysis is not impressive unless it is directly backed up with physical testing. Stuff that clearly shows that the basics of your system are very well thought out."

I paused briefly for other questions before moving on, "Also, as some of you might remember I took the liberty of applying us for several of the awards they are giving out at this thing, so some of you have presentations to give."

"How many did you apply for?" Andy asked.

"Uhhh... all of them I think. Powertrain award, body design award, composites award, they even have a design report award this year. I applied for every award at both competitions."

"Nice," Kyle said with a smile.

"Hey, why not." I said shrugging my shoulders, "We have a good shot at some of them. They all have money that comes with the award and the team always needs money. And I always need résumé padding. So Chris, you've got to present for the powertrain system award, and I'm going to be really upset if we get second place behind Wisconsin again."

Bobby corrected my use of the unmodified university name, "Fucking Wisconsin."

"Right, fucking Wisconsin. Also, unless there are any objections, I'm going to eBay some advertising space on our car. We have some room left on one of the wings, and I figured we could fill it out and make some money in the process."

"How much money do you expect to get from that?" Chris asked in his serious voice.

"Not much, mostly I just think it would be amusing to put advertising space on eBay. I mean if Texas put up advertising space for auction, wouldn't you give a hundred to have a sticker that says 'Texas Sucks'?"

Bobby said, "You realize that means Texas could put a sticker on our car that says 'Oklahoma Sucks'."

"Yea, I'm okay with that. As long as there's monetary compensation. Soooo... any objections?"

"Can we bid on it?" Ricky asked.

"Yea, I don't care."

"Good, 'cause I got twenty bucks for a Ricky sticker."

"Yea, well I got twenty one for a Matt sticker. So there you have it, we're up to twenty one dollars and we haven't even started the auction yet."

Ricky, holding up two fingers shot back "Well I got twenty two dollars."

I glared at Ricky, "We'll settle this on the internet." I moved on to the car, "Okay, we have a few minor things left to do..."

I went through the list of things that needed to be finished up on the car, nothing major, just some things to make it a little more perfect. Bobby finished up the meeting with a reminder that everyone's system design report was due before competition. The system design reports would be assembled into our internal design book, a 200 page legacy that was mainly a way for us to transfer knowledge from year to year. It was also important to have for the design event, because it was a common question for the judges to ask how we were carrying on information.

Bobby concluded the meeting and then added "Oh, and this Saturday we are having a practice competition where we are running the car through all of the events. Feel free to come out and help if you have time. Of course the drivers have to be there."

"I actually have one last thing." I announced, just as everyone was starting to get up. "At the beginning of the year, I started this Mr. McCoy's wall of starts and cars thing, and you've all amassed a lot of stars and cars. I said I'd give a big cookie to the winner. Well, it was just like elementary school, childish and corny, and just like elementary school, everybody is a winner!" There was a collective response of grumbled laughter as I pulled out a giant cookie from behind the podium, along with several paper plates and napkins.

I smiled and said, "Yea, I know. There should be enough slices for everyone, so dig in!"

Our practice event was just to make sure we didn't have some fatal flaw in our car somewhere. We kept track of times, but it was hard to do any comparisons with previous competition times because the track surface was so much different. It's surprising to find out how much difference two similar asphalt surfaces can make on a lap time. We were still a little worried about the cooling system because it was hard to know how much the wings would block the airflow to the radiators on track. So we opted for the highly controlled drive-it-around-and-see-if-it-works test method. Driving the practice endurance event, we didn't see the temperature get much above 220° which is well within the comfortable range of the engine. Detroit would likely have colder weather, so it wouldn't be a problem there. California would be hotter, but we would have to test again before that competition when we could do a test in hotter weather.

After the test, we prepared the car for its trip to Detroit. We added a sticker to our car; someone that Wes knew through an internet forum had won the eBay auction for the sticker and wanted it to say "It's all about Chuck!"

Four days later, we loaded the car into the trailer along with all our tool boxes, extra material, spare parts, everything we thought we might need and more than a few things we were pretty sure we wouldn't need. The competition week was always full of surprises, and we were never prepared for all of them.

We spent all evening packing, and most of us were there until 1am. I showed up at the shop the next morning at 5am, ready to leave. I was so tired. It would be like that for the rest of the week and I knew it. I was hoping someone had brought donuts, but no luck. The trailer was mostly packed and ready to go, we just had to load a few more things and get on the road. Ricky had borrowed his mom's minivan, which was parked out front with the rear hatch open. I put my backpack in and pushed it up against the back seat.

"You have all your spare parts and stuff packed?" Trey asked, walking up behind me.

Trey had been at the shop the previous night and was still there when I left.

"I gotta grab a couple more things. What time did you leave last night?"

"I didn't. Well, I left about thirty minutes ago to pick up some stuff from home, but I haven't been to sleep yet."

"You're not driving are you?" I asked.

"No, I'm riding with Bobby."

"Good, last time you almost killed us all," I said with a smile.

Two years before, Trey had stayed up all night packing and briefly fell asleep while driving. Before we had even left Oklahoma City, I was jostled out of my deeply focused reading by the car jerking to the right and Trey gasping, "Oh Shit! I just fell asleep!"

We didn't think much of it at the time, but when we got to the competition we found out that the team from Montana State had lost control of their tow van and collided with a truck, killing three students. It was a sobering reminder that, even with all the racing we did, the most dangerous part of the competition was the drive there. Phil took safety seriously, and since he was driving the trailer, we could count on the trailer going the speed limit. Although it was frustrating to drive 65 MPH behind the trailer, we appreciated his safety.

I went into the shop to pack up my last few things. I picked up the fragile carbon fiber wing supports and placed them carefully behind the back seat of the tow truck along with the poster boards.

"Phil, do you know who is sitting in the back seat of the truck?" I asked.

"I don't know, maybe no one."

"Okay, well if anyone does, make sure they know that the wing supports are behind the seat and they are like glass."

He nodded.

"Seriously. They can't lean the seat back. At all."

He opened the door, and pulled the ice chest over to that side. "There, now nobody will sit there," He said with his usual smile and laugh.

"Looks good to me."

We finished securing everything in the trailer and left right on schedule. As we were driving, Ricky cranked up the volume to a Beastie Boys song on the radio and we all sang along.

"No! Sleep! Till Brooklyn!... No! Sleep! Till Brooklyn!"

Two hours later everyone in the van was asleep except Ricky.

11

Detroit

"To achieve anything in this game you must be prepared to dabble in the boundary of disaster." – Sterling Moss

Detroit is… depressing. It seems like the sun never comes out, and there are never really clouds. It's like a dome of grey sadness is covering the whole city. I'm sure it's more prominent in the rainy season; we always arrived right at the end of spring. Maybe it's nicer in the summer and fall? It was one of those things that I just didn't really think about until I got there. So when I woke up in the back of the minivan thirty miles from the competition, the excitement of being so close to the track was short-lived and overcome by the realization of an impending week of ominous weather.

We pulled into the paddock and immediately set up our area: a heavy duty awning frame overhanging from the side of our converted horse trailer, draped over with two blue tarps. The tarps had to be taped to the top of the trailer so the rain wouldn't leak between the two. It wasn't the Beverly Hilton, but it did its job well.

The paddock area was a large gravel parking lot lined with

dense trees on two sides. The other two sides opened up to a sprawling competition area. There were a few pre-fab metal buildings that housed tech inspection and the static events, and a large paved area that would be made into courses for the dynamic events with the help of a few thousand orange cones.

One of the first things teams had to do at the competition, after setting up their paddock, was to go through technical inspection. In fact, the cars could not even be started until they completed the first part of tech. While there was always enough time for everyone to go through tech, we chose to plan conservatively; we sent the car and a couple representatives to wait in line before we even finished unpacking. I stayed with the trailer and helped set everything up.

We rearranged the toolboxes so that we could use the trailer as a small workshop. There was a small workbench with a pegboard mounted on the wall behind it. The workbench would soon be cluttered with random crap and be useless as a workbench. The pegboard was never used for anything. We put the toolboxes on either side of the trailer, and everything else was stored in the gooseneck or outside underneath it. This gave us enough room to wheel in the car at night without much effort.

"We won the race to tech! We're first in line," Kyle said walking towards the paddock with his index finger proudly displaying our #1 status.

Wes responded, "Too bad they don't give points for that."

We spent the next few hours walking around looking at other teams' cars. This was the only relaxing part of the week. Everyone went from paddock to paddock with cameras and friendly smiles, taking pictures of cars and asking questions about new designs. Nobody cared if other teams took pictures of their car, but everyone always asked. We had a couple designated picture takers, collecting as many photos as possible. They were invaluable in designing the next year's car; when designing any component you could look at last year's pictures and find the best aspects of all the designs. There were some really clever ideas out there, and we took as many as we could. As the day went on, the line at tech got longer and the cameras and questions migrated over there.

While passing by the University of Missouri-Rolla's paddock, I was stopped by one of their aerodynamics guys. They were one of the other five or so teams with wings.

"I noticed on your car that the end plates for your rear wings extend past the rear tires," he said, "That's against the rules and you'll probably get called out on it in tech inspection."

"Yea, they were designed when we had the Hoosier tires. We switched to Goodyear tires, which are smaller. It's only about half an inch, so we figure we might be able to get away with it."

He shook his head, "Not a chance, they check that on us every year. It's one of the first things they check."

"Eh, we can fix it in just a few minutes with a saw and some scissors, so I think we're just going to try. Thanks for the heads up though."

"Hey, no problem. So who is Chuck?" he asked.

"What?" I asked, confused.

"You have a sticker on your car that says 'It's all about Chuck'."

"Oh. I don't know. Someone from a message board that we advertised the auction on. No one really knows him. But he had a hundred and twelve dollars, so it's all about Chuck!"

"Cool. Well good luck. Nice push-bar, by the way!" he said, walking away.

I arrived back at the tech line just as the car was going into the second part of the inspection. The first part was just to check the required fireproof suits, gloves, and helmet, as well as the two fire extinguishers that we were required to have. The second part was the fun part, where two or three guys would crawl all over our car with a rule book nitpicking every part. That part always started with the tallest driver, Kyle in our case, showing that he could get out of the car in less than five seconds. Fully buckled in with his suit and helmet on, the driver had to be able to exit the car and have both feet flat on the ground in five seconds or less. Kyle and I planned that he would remove the steering wheel, throw it to me, and then I would get out of the way. We didn't want to hurt the steering wheel, and it was precious seconds wasted setting it down carefully.

"Are you ready?" the tech inspector asked, holding a stopwatch intensely.

Kyle nodded.

"Go!"

In less than a second, the steering wheel came flying at me. I snatched it out of the air and stepped back. Kyle was standing in my place almost immediately.

"Whoa, three and a half seconds," the tech inspector said surprised, "I hope you guys are that fast on the race track."

It was a commonly held belief that the tech inspectors always found two things wrong with every car, and so we had purposefully left two small things wrong for them to find, and had the parts to fix them conveniently waiting in the trailer. They found them, we sent team members back to the trailer to "make" the repairs, and we were done.

I didn't buy this two-things-wrong-in-tech idea; I thought they were just very thorough. They had to be; when it came to the balance between safety and speed, we could be like unsupervised kids in very dangerous candy stores. Still, it didn't hurt to have the two things "fixed" during tech.

All of the tech inspectors enjoyed our fabulous push-bar, and when the tech inspectors are happy, things go well. Unfortunately, they also noticed that our wing end plates were not quite rules compliant. I started to argue that the end plates were not part of the aerodynamic device and weren't covered by that rule, but I didn't get two words in before I was interrupted with:

"Don't even try to argue that with me, you can fix them or

remove them."

Mike Royce, the head tech inspector, personally checked all aerodynamics. He was fair to everyone and a nice enough guy, but being head of tech made him really hard to like. Using a rope with a washer hanging on the end, I drew a line on each endplate where they needed to be trimmed. I took them off the car and back to the paddock to be fixed while the tech inspectors finished with the rest of the car. I returned, carrying the modified end plates with a few minutes to spare. I found Mr. Royce, and we got his approval for the new modifications.

As Wes and Brett attached the push-bar to the car Mr. Royce said, "Oh that is fantastic!"

One of our friends from the University of Kansas said, from across the room, "Do the bell!"

Wes pushed the bell lever which gave a "Ching! Ching!" for everyone's enjoyment.

Kyle's parents, Lee and Michelle, had flown in that day and were at the trailer when we got back from initial tech. For the past few years, they had come to the competitions and were always a great help. They had loaned us radios so that the people with the car could communicate with the rest of us without someone having to run back and forth all day.

After initial tech, there was tilt, brake, and noise. The tilt tech was done with a big metal contraption that tilted the car to the side. At 45 degrees, the car couldn't have any fluids leaking. At 60 degrees, the car couldn't tip over. It's pretty spectacular seeing a car tilted to 60 degrees without falling over. Not to be tried with your Ford Explorer. It was required that the tallest driver be in the car during this, and Kyle absolutely hated doing it. His first year, the tire slipped off the table and the car almost crashed down on the ground with him in it. Since then, we carried a board with us that we placed on the table along the side of the wheels to prevent any misadventures.

After each round of tech inspection, the teams would get a sticker indicating that they passed that part of tech. There were three stickers, one for the initial technical inspection, one for the tilt table, and one for the noise and brake test. We had two down and one to go.

Brake was just accelerating and braking in a short distance, in which you had to lock up all four wheels. Noise was an inspection of the car's exhaust noise level, measured half a meter from the exhaust outlet. The loudest allowed was 110 decibels. 110db is about like a concert or a loud nightclub, but the noise level drops off a lot as you move away from the car.

Noise came first. We had a meter to test this, but we were never sure if it was accurate, or even repeatable.

It turns out it wasn't, because we didn't pass the first time. We had to bring the car back to the paddock and try to modify the muffler.

"How loud was it?" I asked.

"One fourteen," Bobby said without looking at me.

I stood silently stunned for a moment waiting for him to say "just kidding," then I said, "No shit!?"

"No shit."

"Son of a bitch. That's gonna take some serious modifications to bring that down," I said shaking my head. The decibel scale is logarithmic, so the difference between 110 and 114 was a lot more than the difference between 106 and 110. In other words, we had a long way to go.

Back at the paddock we opened the muffler up and modified it by lengthening it slightly and adding new packing. Back at the noise test area, Bobby, Kyle and Ricky tried again. If they passed noise, they would go directly to brake test in the same area. After a quick test, we saw them headed back our way.

"One fourteen," Bobby said.

It was going to be a long day.

"But we got our decibel meter calibrated." Chris added, holding up our cheap Radio Shack decibel meter.

Wes and Brett pushed the car under the awning and started to take the muffler off. I pulled out the red tub of exhaust parts and brought it out of the trailer.

"Let's try the big muffler," I said.

"What's that going to do to the power?" Kyle said.

"Make it go down," I said matter-of-factly.

"It doesn't matter how much power we have if we don't run," Ricky said, pointing out the obvious.

"Well, if we don't have to use a muffler that kills all our power, then—" Kyle started to argue.

Ricky interrupted, "I know, I know, I'm just saying we need to pass."

"This muffler isn't going to fit. The exhaust is too close to the suspension," I said.

"Can we extend the outlet?" Chris asked.

"If we had an extension piece," I answered, "You know we could just repack the old muffler all the—"

"We just tried that," Ricky interrupted.

"Yea, I know," I snapped, "let me finish. We could repack it all the way, and get rid of the resonance chamber. Everyone else is using a fully packed muffler and it works for them. Well, most of them."

Lee leaned in and said, "If you guys need an exhaust extension, I can go get one."

"Yea. I think we're going to need the big muffler," Kyle said.

I shrugged my shoulders, "We can try to get something to work here while you're getting the parts for the big muffler."

Kyle and Chris made a short list of pieces for Lee to pick up while Ricky and I pieced together an extended core. After a few minutes, we put it on and tested it. The meter read 113; we were going in the right direction, at least. Someone suggested using less packing so that it would absorb more sound, so we took off the muffler, drilled out the rivets, lightly packed it, and put it back together. 115, the wrong direction. The obvious path was to use more packing and make it denser, so we took apart the muffler again and wrapped it as tightly as we could. We realized that we didn't have enough packing to fully pack it, so we wrapped it tightly around the core, and then wrapped the whole thing with stainless steel wire to get the packing tightly around the core. We tested again and we were just below 113.

"Hey Kyle," I said, "can you call your dad and ask him to pick up some more packing while he's out."

"How much?"

"Like... a lot."

While we were packing and repacking, Phil was building another muffler, adding chambers and baskets, using parts that were never designed to be in an exhaust system. We were getting desperate.

Ricky made a suggestion, "Let's see if we damp the sound coming from the engine with some of that rubber floor padding. Just see if it helps."

"What... how are we going to do that on track?" I asked argumentatively.

"I don't care, let's just see if it helps," he snapped back, "we can figure that out later, we need to know where the—"

I interrupted "Okay, okay, alright, we'll try it." I was trying to avoid a pointless argument, but I was still using an argumentative tone. Also, I started the argument. I was getting frustrated.

Damping the sound coming from the engine itself helped a little, but not enough. We tried Phil's muffler with some success, and we retarded the ignition timing which helped a little. Lee returned with the exhaust extension and more padding. We re-packed the old muffler with the new packing and got down to 112. We were trying to do things that would have the smallest effect on power, but that was rapidly becoming more difficult.

Kyle, clearly frustrated, said, "I don't get it, Kansas has a through packed muffler and the same engine as us, and they're at one hundred eight."

I responded dishearteningly, "Yea, but they do have a different exhaust layout; our pipes all collect at the same place, and they have a four-into-two-into-one setup."

"You know," Ricky said, "they have their old car here. We could put their old exhaust headers on our car."

"Yea but…" I started and then stopped myself. I was initially reluctant, but then thought about how we were running out of time. The next option was to use two mufflers, which would create a weight and packaging nightmare. I continued, "…I'm up for it if you are."

Ricky shrugged his shoulders and said, "Let's do it."

I started to grab some tools when Bobby said, "Guys we have the design event in fifteen minutes, we need to start getting ready for that."

"Yea," I said looking at the clock on my phone, "Okay, let's get a muffler back on so we can have a complete car. Whichever one is easiest to get on and still looks decent."

I grabbed our poster boards and a few other things and headed towards the design area.

Wes and Brett pushed the car into the designated spot just as I finished setting up the boards. Design preliminaries were a chance to show the judges that there was nothing wrong with the car, and that we more or less knew what we were talking about. Every year I had been on the team someone had taken the opportunity to ask the judges for some feedback and advice. One of the things they all said was to have a brief introduction to the car and the designers of the major systems. Following this advice, I had prepared a speech and had practiced it so that I could convey the basic information and the who's who of the Oklahoma team.

The judges walked over to our car and everyone was ready to go.

"Okay, we are the University of Oklahoma, this is—"

One of the judges, presumably the head judge for that queue, interrupted me, "This is the way we're going to do this, we're all going to pair up and talk to each of you about your systems. Who is your engine designer?"

He started listing off systems and our team members stepped forward when appropriate. After everyone was accounted for, he continued, "I'm going to start with aerodynamics and then move to engines, so let's get started."

He walked over to me and asked, "Where is the center of pressure?"

Easy question. I answered, "It starts about three inches in front of the center of gravity and moves backwards to about an inch behind at —"

"Did you test that?"

"Yes, we have wing testing in the wind tunnel backed up with—"

"So you didn't do full car testing?"

"No we did a full car CFD analysis, but our wind tunnel testing was with the individual—"

He shook his head while I was talking and interrupted me again, "Now how did you choose the mounting points for the wing back here?" he asked, pointing at the rear wing.

I paused and tried to think of why he would ask that. I couldn't think of anything, so I just answered honestly, "It packages well. Those are easy locations for installing the—"

"You didn't do any testing to see the effect of mounting locations for the rear wing?"

I answered quickly, hoping to complete a sentence before being interrupted, "No, the wing itself is as far back as the rules allow, they are up to the rear edge of the rear tires."

"I'm not talking about the wing, I'm talking about the mounts for the wing. Why didn't you put them farther back?"

I thought again. Why would I do that? Where is he going with this? "No, because if the wing stays in the same spot, the net force on the tires is the same, independent of the chassis mount-ing location."

He shook his head, "Think about what happens if you have the mounts for the wings farther back."

I stood silently, looking confused.

"You get a moment acting at the mount," He looked at me with an expression like he expected a light bulb to go on. But none.

"Think about it," He walked off, over to Chris to talk about the engine, presumably.

I did think about it. How could a moment just be created without being countered by another moment or force at the other wing mount? What was this guy talking about? After the end of our allotted 20 minutes, we thanked the judges and pushed our car back to the paddock. On our walk back, I talked with the other teammates and got my answer.

"He was asking why we didn't have all four corners identi-cal," Chris said with a confused look.

"We have all four corners identical, from hub bearings out," Kyle insisted.

"He wanted all the A-arms to be the same, too. So that we would only need one set of spares."

"That's retarded," Ricky said. Ricky was not a suspension person, and neither was I, but both of us, and everyone on the team knew that the tradeoff of convenience for the loss in on-track performance would be a ludicrous decision to make. It would be, as Ricky put it, retarded.

"Wait, why was he asking you about the suspension, wasn't he the engine guy?" Bobby asked.

I jumped in before Chris responded, "Yea, he was ask-ing me some stupid question about the aero, too. About why we didn't have the wing mounts farther back on the frame."

The more I thought about it, the angrier I got. I felt like we just got cheated out of a decent design event by an arrogant jackass who didn't really understand the systems he was asking about. He was an engine engineer for some company that made powertrain parts, so why was he asking asinine questions about the aerodynamics and suspension.

Wes, as usual, stifled my bitchy mood with a positive note, "One of the other judges really liked your shifter and clutch. He liked that you could clutch or shift up or down with either hand."

We had push button shifting that I had setup so that the driver could clutch, shift up, or shift down with either hand. It didn't seem particularly useful, but I knew it would be something the judges would like. Unfortunately, I didn't think it would be enough to overcome all the "shortfalls" we had elsewhere.

In any case, we had other problems to worry about.

As soon as I got back to the trailer I asked, "Are we still going to try the Kansas exhaust?"

"Yea, we need to get this done now," Kyle responded, clearly frustrated at not only our design event but also our exhaust dilemma.

I grabbed a small assortment of wrenches while Kyle went to tell Kansas of our plans. They wouldn't mind; in fact, they would be happy to help. The University of Kansas was one of the nicest teams there. Also, we were friends with most of them so they would have let us use anything they had, so long as they didn't need it. While we were disassembling Kansas's old car, Chris and some other team members were installing the large muffler. They tested it, and it was an encouraging 111db. They disassembled and repacked it again, but it was still 111. Someone suggested adding some baffling in it, like Phil had done with his other muffler. So they started taking apart mufflers and collecting pieces for a mega-muffler.

I was uncomfortably underneath Kansas's 2006 car with my arm reaching up into the exhaust area, scraping my knuckles on parts of the car I couldn't even see.

"Are you enjoying our packaging arrangement?" I looked up to see Erich, one of Kansas's team members.

"This has to be the most difficult exhaust system ever to remove," I said, "How the hell do you take it off?"

"We usually pull the engine out first," He said with a smile.

I continued removing bolts, using the method of balancing a wrench between two fingers, blindly slipping it over the bolt, turning the bolt a tiny bit, and repeating. For good measure, I sprinkled in some dropped wrenches and a few dejected "goddamnit"s.

Bobby walked up, "They're going to close the paddock in a few minutes. We need to figure out what to do."

The teams were allowed to take the cars with them at night, as long as they went through all tech inspections again the next morning. We were done with all but two, so it would have been a huge hassle. However, we needed a solution.

"I think we should take it home," I suggested, sitting up.

"You mean… back to Oklahoma?" Kyle responded, sounding like someone just shot his dog.

"No, no no no no. No, I mean back to the hotel to get the muffler thing figured out," I responded, clarifying my use of the word 'home'. "No way we're giving up. Not until they make us. And maybe not even then."

Ricky said, "I don't think we can do sound tests in the hotel parking lot at night. If we take it off site, we have to go through all the other tech inspections again, and we only have until noon tomorrow before the skidpad and acceleration events close. I think we have a better shot fixing it tomorrow morning."

"That's true," I replied calmly, "as long as we get here as soon as we absolutely can, and we have to realize that we *need* this done no later than eleven."

"Alright, I'll go tell everyone else," Bobby said.

The other team members packed everything up and got ready to leave. I finished removing the last few bolts on the exhaust just before we left.

After what seemed more like a nap than actual sleep, we were on our way back to the competition. Bobby, Kyle, Chris and I went ahead of everyone else to make sure we were there as soon as the gates opened. We immediately started installing Kansas's old exhaust headers on our car. Lee and Michelle arrived soon after us, followed by the rest of the team.

As we finished installing the exhaust, Chris was frantically looking for the muffler that some of the team had been working on the night before. We grabbed a random muffler and tried it.

"One eleven," Bobby said, "That one was one fourteen before."

Kyle made a suggestion, "Put the muffler from Kansas on the end of ours, so the two of them are in a row."

Wes put on a pair of gloves and held the muffler to the outlet of the other muffler and Kyle started the car.

"One oh nine, that's it." Bobby said understatedly.

Chris shouted, "Where is the other goddamn muffler!"

"We're out of time," Bobby said, "we have to do something. We have to use two mufflers if we don't have anything else."

Now we just had to figure out how to package it. We couldn't just run them in line, for one it was against the rules to have the exhaust extend back that far, but also because it would be swinging around, knocking over every other cone on the track. We drew up some crude plans for two mufflers doubling back on each other behind the car, and sent Lee out on another parts run. Fortunately, he knew where all the good exhaust parts places were, since he had so much experience the previous day. Still, our time was running out and we were all hoping God could somehow miracle him to the nearest parts store and back.

After he left, we tried a few other things in hopes that something, anything, would pass. We got about half a decibel lower with everything we tried.

"Two hours Phil!" I looked up to see Chris picking a muffler up out of the bed of the truck. He was yelling and waving the muffler at Phil, "Two fucking hours! I asked you if it was in the truck and you said no! Two fucking hours!"

It was getting late in the morning, and everyone was tense, tired, and angry. Although the attitude wasn't completely alien for Chris, he usually refrained from outright yelling.

Chris started pulling off the other muffler. We all kept our distance. Once he had the new one firmly clamped on, he told Kyle to start the car. I leaned over Bobby's shoulder as he read the meter.

Bobby announced, "One oh nine! That's it. Let's go!"

He barely got the words out of his mouth before everything was ready to go and they were on their way to the noise tech area, pushing the car at a jogging pace.

So with exhaust pipes made in Kansas and a muffler built from pretty much every random piece of muffler we had, including a few extra pieces of whatever the hell was laying near the table, the car was off to try to pass sound. We all stood by the fence watching with what seemed like the last bit of hope we had left for the sound event. If this didn't work, we would go back to the trailer and weld together two or three mufflers, severely decreasing our power and likely causing us to miss an event. The engine revved up as the sound inspector stood just behind the car. After a couple seconds, he picked up our tech sheet and wrote something on it.

"Did...did we pass?" I asked.

The inspector handed Bobby the sheet, and they started towards the brake tech area.

"We...we passed. Did we pass? I think we passed!" Relief and relaxation is not a feeling that comes often at the race track, but I sure felt it then. After all that, we had finally passed sound. Now all we had to do was pass the simple brake test and we would be on our way to the races.

Kyle revved up the engine and started to ease the clutch out when "Pop!" the internals of the muffler exploded backwards and scattered for 30 feet behind the car. We had packed so much crap into the muffler that it had built up enough pressure to blow it apart. The core had been wrapped with white packing which feathered off as it zoomed across the pavement. It looked like someone had shot a chicken out of a cannon. It probably would have been hysterically funny if the circumstances were different, but instead it was just really funny.

"Get the rivet gun and a drill so we can add some more rivets," came over the radio authoritatively. They picked up all the muffler innards, pushed the car back outside the dynamic area and reassembled our frankenmuff when the tools arrived.

"Here, you drill five or six holes and I'll start adding rivets," Kyle said anxiously, handing the drill to Ricky.

"Well, at least we passed sound," I said.

"Nope, we have to do it again," Ricky said, still focused on his drilling task.

In calm disbelief, I asked, "Say what now?"

"The guy said we have to do it again after we fix the muffler."

"Well son of a bitch."

So much for that feeling of relief.

Five minutes later, they were back to recheck sound and we were all sitting there thinking the same thing: now we're going to be too loud again.

Fortunately, they had gotten all the feathers back in the frankenmuff and passed sound. They pushed the car back to the brake tech area, and I stood tightly gripping the fence and hoping that there wouldn't be a repeat of the chicken cannon.

Kyle revved up the engine and eased out the clutch. He accelerated for about two seconds and slammed on the brakes. All four inspectors raised their hands, indicating that all four wheels had locked up. Tech inspection was finally over. We had spent the last two days stressing out, running back and forth, yelling at each other, and frantically trying to make things work. And we hadn't even started racing yet. Now all we had to do was get Ricky in the car, get in line, drive the skidpad, drive the acceleration event, switch drivers, and repeat. All in less than an hour. Oddly, that all seemed pretty easy.

I turned around to see Lee, standing with a bag full of exhaust parts. He laughed, held up the bag, and said, "I got 'em!"

12

The Races

"To finish first, you must first finish." – Rick Mears

We had found out that we didn't make it to semi-finals in the design event. In fact, no one in our design queue made it, including Western Australia who was an excellent team and a perennial favorite for design finals. We took this as more evidence that our design queue was run by an arrogant and retarded monkey who was shaved and clothed to look like an engineer. Still, all our angry ranting and fuming wouldn't change the result. Perhaps more importantly, we had other pressing issues at hand.

Ricky, Chris, and Bobby were rushing to the skidpad and acceleration events. There were less than 45 minutes until the events closed. I was about to pick up the radio and tell them to only do one skidpad and one acceleration so we could make sure to get points in each before going back for our second driver. But I assumed that they would figure that one out on their own, and my insistence over the radio would just cause frustration.

Ricky was dressed in his fireproof driving suit and in the car ready to go.

Lee had his headset on and radioed into Kyle "The line for skidpad looks pretty short right now, you might want to do that first."

With no response, Kyle and Bobby pushed the car up to the line for skidpad. The event area was pretty far away from the spectator area; we could barely see the timing marquee for skidpad. Fortunately, we had chosen the same radio channel as one of the other teams and they had a person relaying times back to their trailer, so we knew how fast everyone was going.

"Do you know what time it is?" I asked Lee.

"Eleven twenty one," he said.

My heart was beating fast as I stood nervously watching them wait in line. The other cars were systematically going through the event, and after only a few minutes, we were next.

"What's everyone running in the skidpad?" Wes asked.

"Five point four is about average for the good teams," I said, still staring intently at our car.

As though that was his queue, Ricky took off around the skidpad, twice clockwise then twice counterclockwise. I squinted at the timing board but couldn't tell what it said.

Dr. Siddique was standing next to me with binoculars and announced, "It looks like a five point four."

"Not bad, not bad," I said quietly. "I think we're probably going to make all of our runs."

Bobby and Kyle pushed Ricky to the acceleration event and, after a short wait in line, he ran a 4.47. Not a great time, but acceleration wasn't our best event. After his two runs, Bobby got in the car and posted a time of 4.318. The best time is the only one that counts, and that time put us in with the bubble teams, around 25ᵗʰ. They pushed the car back to the skidpad to give our second driver, Chris, his shot. With only ten minutes to go before the two events closed, Chris posted a 5.173, a time which put us in fourth place in that event.

We pushed the car back to the trailer, feeling relieved but too worn out to be excited. Mitch, Gloria, and Michelle had set up lunch for us. Between Kyle's parents, our two faculty advisors, and Gloria, who we often called the team mom, we had a full blown support crew. It was like having a catering company, an equipment supplier, and a parts delivery service. It was clear that we wouldn't have made the events without their help, and we wouldn't have been as competitive as we were without their help all year long.

After lunch, Kyle and I would go out and run the auto-cross event. I was exceptionally nervous, so I decided to sit alone under the gooseneck of the trailer. I looked over the autocross map carefully. I closed my eyes and imagined each corner. *Enter fast, brake late, and carry speed through the big corners. Oscil-late the slaloms quickly but not recklessly; trust the car but don't throw the car.*

While I was going over the course map, I overheard Chris apologizing to Phil for yelling at him earlier. Phil dismissed it, saying it was water under the bridge, but I knew as well as Chris did that Phil appreciated the apology. And of course Phil knew that Chris was just stressed out and overreacted. They both had made mistakes, but neither of them was going to let stressful cir-cumstances ruin their friendship. I wished I had thought of that.

"Matt, did you get some lunch?" Gloria asked.

I looked up slowly and calmly said, "No, I think I'll wait to eat until after the autocross."

"Okay, I'll make a sandwich and save it for you."

"Thank you," I replied. She didn't ask what I wanted on my sandwich, which meant she could tell that I was trying to be left alone.

After a few minutes I was calm and focused. When it was time, I put on the driving suit and walked with Kyle down to the autocross event. The track was open for walking for a few min-utes before the event started. The map was close but not exact,

141

and even if it was, walking and seeing the track from the driver's perspective before the race was invaluable. Kyle and I discussed how we were going to enter each corner, how we would take the slaloms, and where we needed to be careful.

I was the first driver, so when the car showed up, I got in and got ready.

"We need to wait until a few cars go out and sweep the dust off the track," Kyle said as I was buckling up, "You might not want to get in just yet."

"I'd rather be in the car," I responded quietly. I could focus on imagining the track better while sitting in the car. Also, it was a very comfortable car; it wouldn't make a bad lawn chair. We waited beside the queue for a while, and then Ricky and Kyle pushed the car to the back of a long line that had formed. Each team would have one driver do two laps, and then after all the teams had one driver go, the second drivers would do two laps. I sat quietly for a long time, thinking about the track, going over it in my head, and deciding which areas I was going to test on the first lap.

"Alright, you ready?" Ricky asked.

I nodded. My heart started beating fast again. I nervously pulled the straps on my seatbelt to make sure they were tight. The car in front of us got the green flag and accelerated away. I eased out the clutch and slowly rolled to the starting line. I took deep breaths and focused on the first three corners. I knew that by the time I was through them, I would be fully focused on the race. Deep breath, accelerate for a short distance before slowing and turning left, then immediately right, then a sweeping accelerating left. Deep breath, accelerate for a short distance before slowing and turning left, then immediately right, then a sweeping accelerating left.

The starter waved the green flag. I took a deep breath, eased out the clutch quickly and accelerated into the first corner. I could feel my left leg shaking on the pedal while braking into the first corner. I was too tense, so I took another deep breath and reminded myself that the first run was for warming the tires and practicing the course.

"This is the test run, just the test run. Focus on the car," I said quietly to myself, under the 109 decibel roar of the engine.

I eased off and slid a little through the turn. I stabilized the car for the right turn and by the time I hit the sweeping left I was focused. The car slid through some of the corners; a few of them I did on purpose to get the tires up to temperature, and a few of them were just because I was driving in too fast. There were three big sweepers near the end of the course and the car cornered like it was on rails through them. I pushed harder and the car just stayed in the corners. I pushed the pedal farther and farther on the last sweeper, determined to find the limit.

"Faster, faster! GO GO GO!" I yelled, easing the pedal closer to the floor, searching for the limit of the car. I found it just as I needed to start slowing down, and almost took out a dozen cones. I finished up the last few corners and came across the finish line. Quickly I drove back to the re-run line, trying to get my second lap immediately while the tires were warm. After just a moment, I got the green flag again and I was off. Left, right, sweeping left. There was a little less sliding, but my corners weren't perfect. When I got to the sweepers, I pushed it all the way from the beginning. I didn't get the transitions between the sweepers perfect, but it wasn't bad. It wasn't the best I could do, but it was good, top fifteen maybe. As soon as I finished, a guy wearing a Goodyear shirt started measuring the temperature of our tires. I took the opportunity to ask him how I did.

"What was my time?" I asked.

He said something quietly that I couldn't hear over the engine, so I pushed the kill button and asked, "What?"

"Forty seven-seven. Fastest time so far," he said with a smile.

As soon as he said it, Kyle and Ricky came over with the

push-bar.

Ricky announced, "Alright Superfast, fastest time so far!"

I tried to hold back my huge smile, but I'm sure everyone could see it, even with the helmet still on. I wanted to play the calm, cool guy: *I knew that, I expected to be the fastest, I wasn't worried at all.*

But instead I just smiled ear to ear.

The University of Texas Arlington went out shortly after us and put us in second place. Everyone sends out their slow driver first so they can heat up the tires and give advice to the fast driver. I was the second fastest of the slow drivers, not too bad. As all the faster drivers went out, we were still in second until Kyle went out and took first place back. A few cars later UTA's second driver went out and took first once again by just over two tenths of a second.

Second place was fine with me. Kyle was a little upset that he missed first place, but he and the rest of us were still happy. I was much more relaxed, and I knew I could go into the endurance race without being too nervous. My time was fourth overall, behind Kyle and the two UTA drivers. It was clear that we had the car and the drivers that we needed to win.

We packed up for the night and headed back to the hotel to get some sleep before the big race the next day.

As we were setting up our area the next morning, Brett added up all the scores. The design scores would change depending on design finals, but we were firmly in the top five, and not too far behind first. We could expect at least a couple of the top teams to drop out of endurance for some random failure, and so we just needed to drive a little faster than a few of them. We were all surprised that we were doing so well. We expected it in a way, but it was still hard to believe that we were really a top level team with a top level car. As long as any of us had been on the team, a top five finish was a long shot at best, and now we were very close to top three, and we had a real shot at first place.

"Is UTA ahead of us?" I asked.

"No, but it's really close," Bobby answered before Brett said anything.

"Who's in first?" I asked.

Brett just looked at me, smiling.

"Awww, don't tell me it's fucking Wisconsin."

He nodded and laughed, and I shook my head in disappointment.

"Why do you hate Wisconsin?" Wes asked.

"I don't really hate them, it's sort of a rivalry, but more one-sided."

Wes still looked confused.

"Okay, last year I was trying to win the powertrain award, and I knew Wisconsin was going to be tough competition. So I went over to their car and picked out the things that were wrong, so I could point out how they were right on our car and maybe the judges would then realize they were wrong on Wisconsin's car. Turns out all I could find was that their CO_2 bottle for their pneumatic shifter was oversized. So when I was giving my presentation I mentioned how our pneumatic cylinder was small and not the size of Wisconsin. Well, it either didn't help or backfired 'cause they got first place and we got second. From then on it was 'Fucking Wisconsin'. It's a bitter rivalry, like the Hatfields and the McCoys, except Wisconsin doesn't actually know about it."

"So you don't like them because they're better than us?" Wes asked, looking for more clarification.

I shot back indignantly, but with a smile, "They're not better than us, Wes."

"Is that why they're in first—"

I interrupted with an impenetrable argument, "Wes... Okay... Wes..."

Kyle and I got ready to drive while the rest of the team buttoned up some things on the car. They taped up the body panels and fuel cap, just in case, and made sure all the bolts were tight.

The Endurance course was similar to the autocross course in size, but it was setup with different corners and slaloms. Also, instead of driving one lap at a time, we would each drive ten laps. I would drive first, and after my ten laps I would have to pull into the driver change area, turn off the car and get out. Kyle would get in, fasten his seatbelt, start the car and go finish the race. From the moment I took the green flag until Kyle finished, we couldn't work on the car at all; it would have to start under its own power for Kyle and it would have to stay in one piece. It sounds easy enough, but fewer than half the teams finished endurance each year.

When the time came, all the endurance drivers went out to walk the course. Kyle and I again discussed how we would take each corner and where it might be deceptively slow. Going off course in endurance would result in a 20 second penalty, which would kill our chances of winning. Even the extra two seconds for hitting a cone would be unacceptable, so we would have to be very careful on our first couple laps while the tires were still cold. Careful but not slow. Autocross was a challenge to drive as fast as possible without hitting any cones. Endurance was a challenge to not hit any cones, while driving as fast as possible. I needed to stay calm and focused, carefully navigating every corner and improving every lap to perfection.

There would be about five cars on track at any given time. Whenever one team finished their 20 laps, another would be sent out. The starting order was based on the autocross finish times, so everyone on the track would be about the same speed. Occasionally, one car would get close to another, and the car in front would be given a blue and yellow flag indicating that they would need to pull over in one of the designated passing lanes to let the car behind them go by. It was likely that I would pass at least a couple cars, but not very likely that I would get passed. UTA was the only car faster than us, and they were only a fraction of a second faster per lap.

I was calm, or at least calmer than I had been before the autocross. When the car arrived I got strapped in and went through the track in my head again, just like before. Deep breath, immediate left, sharp right, sharp left, short slalom, sweeping right...

A man wearing an orange vest and headphones walked over to the line of racecars and said, "Two minutes! We're going to start in two minutes!"

I took a deep breath and said, "I'm going to warm up the car."

I announced my intentions to see if there were any objections, but nobody said anything. I pressed the arm button which turned on the car, then the start button to start the engine. It started immediately. I expected we wouldn't have a problem restarting when Kyle got in, but it was still a worry because so many teams dropped out of the race for just that reason.

UTA was waved to the starting line, and I followed several feet behind. The starter pointed at their driver, who gave a thumbs up. The starter then waved the green flag and UTA's car shot forward onto the track, spinning tires a little bit and throwing rocks back towards the rest of us. I slowly drove forward to the starting line and waited for the starter. We both watched UTA on track, waiting for them to get some distance.

The starter then pointed at me and waved the green flag.

I took a deep breath and carefully accelerated into the first corner. I put the car in second gear so it would be smoother going through the next two tight corners. As I turned the sharp left right before the first slalom, the car died. I was worried, but assumed I had accidentally bumped the kill button, so I reached up and pressed the arm button. Since the car was still rolling and the engine was still spinning, I didn't need to press the start button, the car just sprang back to life. I had messed up the entrance to the slalom, but it was so slow that it didn't really matter that much.

I focused back on the race, entering the corner after the slalom carefully. We had noticed some dirt and gravel there while we were walking the course. I took the corner slower than I thought I could, just in case, but the car was sticking to the track just fine. I took the next two turns a bit more aggressively. Just as I was focused back on the course, the engine died again.

I pressed the arm button and it sprang back to life once again. I started to get very worried. I was fairly sure I hadn't bumped the kill button twice. Still, I reached over with my right hand and pulled my left hand glove tight just in case my glove was loose and brushing the switch.

The course had two big sweeping corners near the end. That was where we would be able to gain a lot of time over the cars that didn't have aerodynamics. As I was taking those corners, I looked up to see where UTA's car was so I could check again the next lap to see if I was gaining or losing time on them.

I took the last corner before the finish line a little too fast, but I kept it in between the cones and didn't lose much time on the correction. As I was entering the slow tight corners at the beginning of the track, the car died again.

"Oh no. No NO NOOO! What are you doing!" I yelled in my helmet. I pressed the arm button and the car started back up. As I was going through the course, I started to push on the wires underneath the dash, thinking maybe there was a loose connection and the wires were just pulling away from each other in the corners. The second lap it happened five times. By the third lap, the car would die almost immediately after I took my hand off the arm button.

With my left hand resting on the front roll hoop and my thumb planted on the arm button, I steered through the course as best I could. I could shift both up and down with just one hand, but it was difficult to do while steering with the same hand. I left the car in second gear which made the slow corners slower, but since it was very difficult to steer anyway, it didn't really matter. The big sweepers were easy to steer, so I could shift into third gear at the beginning and back into second at the end without much trouble. I was gaining on cars in the big corners, but losing more time on the rest of the track. During my fifth lap, one of the flag stations waived a blue and yellow flag at me. I pulled into the passing lane, and an orange and blue car shot past me: The University of Florida. We were much faster than they were in the autocross event, and yet I was losing so much time that they passed me. I pulled back out onto the track right in front of the sweepers. I was going so much faster than Florida in the sweepers that I almost caught up to them at the end, but after just a few tight corners, they were gone. During the next lap, I was passed by the University of Western Australia.

My steering arm started to get tired. I was leaning into the steering wheel to get more leverage, but I was only getting slower. For the rest of my ten laps, I tried to position myself to get the most steering force. I got the one-lap-to-go signal and then realized Kyle would have to start the car, shift into gear, and drive ten laps with one hand on the arm button.

I crossed the finish line for the last lap and pulled into the driver change area. I unbuckled, jumped out of the car, and said to Kyle "You have to hold the arm button the whole time. It won't run at all without the arm button down." I was trying to speak loud enough for him to hear, but not loud enough for the official to hear in case he wanted to consider it a safety issue and kick us out of the race. Kyle nodded his head and got in the car. When he was finished buckling up, he awkwardly placed his hands so that he could get one hand on the steering wheel pulling in the hand clutch lever, one finger holding the arm button, and one finger pressing the start button. The engine turned over several times, but didn't start. Kyle stopped for a few seconds and tried again, with no luck. He sat for a moment, letting the car cool down before trying again. Still the engine wouldn't start. The car was not designed to be run with the arm button down, obviously. With the button pressed, all the electricity was going through some path that would allow enough for the car to run, but not to start.

"Okay, I'm starting the clock," the official in charge of the driver change announced to us, "You've got two minutes to get it started or that's it."

Kyle pressed the shifter buttons to get the car into neutral, and then tried again. Still the car wouldn't start.

"Tell me when we have ten seconds!" he yelled through his helmet at the official, who nodded in agreement.

We waited for almost two minutes, hoping the car would start after it cooled down for a moment. I stared blankly at the engine, wondering what had happened, wondering if I could get away with doing anything to fix it without the inspector noticing, wondering if I could pee on the engine to cool it down.

"Ten seconds!" the official said, pointing at Kyle.

Kyle held down the start button and the engine cycled slowly. It didn't even try to start.

After our ten seconds, the official shook his head and said, "Sorry guys."

We were finished. No big trophy, no big #1 to put on our résumés. What was, just hours before, a team with a top car and a real chance at a great finish had disappeared, replaced by a broken car and a handful of dejected kids with barely the energy to push the car back to the paddock.

I stood quietly for a moment in a daze. Florida's driver pulled in to the driver change area, and one of their students bumped me as he ran over to the car. I snapped out of my daze, stepped out of their way, and walked back to the trailer as the competition went on without us.

Ricky knew what the problem was before we even got back. He took off the kick panel and plugged in a wire that had become disconnected. He pressed the start button and it fired right up. The brake over-travel switch wiring had come loose and the car had shut down, thinking the brakes had failed. It had slowly worked its way loose on track. The fact that the car would run with the arm button pressed was a coincidence, a side-effect of the way the car was wired.

"Well," Bobby said with an obviously artificial upbeat voice, "let's go talk to the design judges and see what we can do better at in California."

The team slowly pushed the car over to the design area. There were always some judges giving advice on making the car better and doing better in the design event. Design wasn't where most of the points were, but it was where the prestige was. And the jobs; most of the judges were from the auto industry and looking for new engineers.

Bobby found one of the senior judges, Claude Rouelle. Claude owned a consulting company, Optimum G, which sold software for suspension analysis. He was one of the most well respected engineers at the event, and well known worldwide in racecar engineering. He spoke in a Belgian accent that would easily be mistaken for a French accent by someone who didn't already know otherwise. All the newer members stood out front eagerly waiting to hear what he had to say about our car. I stood quietly in the back, behind the car, still disappointed at our failure.

The design event was a professional setting, almost like a job interview, so we were a little taken back with his fist comment.

"You guys were fucking flying out there."

He must have noticed our reaction, because he followed up by casually saying, "I'm speaking American, now."

He looked at our car and made some comments, almost everything positive. Bobby asked him why we didn't get a better score in design.

"I don't know," He answered, "I wasn't in your queue so I don't know what questions were asked or how you answered them. Your car is good, though. Excellent, I would say."

He answered some more questions, made some jokes, and imparted some wisdom.

"Racecar engineers can make good money, not rich, but good. Remember, a man can have many beds, but he can only sleep in one of them," We all nodded silently, "Of course, he can have many women in that bed."

We all laughed, and hoped he would be in our design queue in California.

By the time the awards ceremony started we had all relaxed a little and were in a better mood. We knew we still had California, and we expected to take home at least one award. We actually took home five: the design and communications award for our design report, Hoosier Tire autocross award for Kyle's top three autocross finish, a body design award, a composites award, and second place powertrain award.

Not only did Wisconsin beat us in the powertrain award again, they got first place in design and first place overall.

Fucking Wisconsin.

As it turned out, we were 28th. Going back through the points, we found out that if Kyle had a chance to finish the endurance at the pace I was going, even as slow as I was with one hand, we would have finished second overall. If we had done better in design and had two hands for the endurance...

Well, there were half a dozen teams that could have won if this had happened or that hadn't happened. There always were, every year. We did well, and we would do better in California.

I think our proudest accomplishment at the Detroit event was our push-bar. Everyone loved it so much that the organizers of the event created a new award, Most Outstanding Achievement in Push-Bar. We got an honorable mention, and they made it an actual award the next year, with a trophy and some money to go along with the honor.

13

More Final Touches

"It's not always possible to be the best, but it is always possible to improve your own performance." – Jackie Stewart

On the way home from Detroit, we had all agreed to take a week off, but I had no intention of leaving my last month on the race team three quarters full. I took a couple days off and then gave into the urge. I decided that I needed to clean up the rear wing, and it needed to be done sooner rather than later. When I cut the endplates shorter to pass tech inspection, I had left them a little rough looking. I had no expectation of anyone else being at the shop, but I was pleasantly surprised to see Kyle there when I arrived.

He was bolting on a new muffler, the same brand that Kansas used.

"I thought we were taking a week off," I said, walking over to the car.

He shrugged his shoulders, "This needs to be done. Chris was here earlier playing with the ignition timing and we had it down to one oh nine with this muffler."

"Heh, did anyone take the week off?" I grabbed the allen

wrench set and started taking the rear wing end plates off.

"I figured you would be looking for a job this week," Kyle said, "I think you're the only one with nothing lined up."

I hesitated for a moment and then replied with my big news, "I got a call from Claude."

"Claude? Claude Rouelle?" He asked, eyes wide.

"Yea, he wants to talk to me at the competition in Cali about a job in the engine department of the team he's working for now."

"Dude, that is... that is really cool."

"Yea. I didn't really expect to hear from him, I just sent him my résumé because I had printed thirty of them and had some extra," I smiled.

"I think it would be a great opportunity to work for him. You should definitely see him at the competition."

"Oh, I have every intention of trying to get this job. I just didn't think I had a chance. That just goes to show you... something. Aim high, or... I'm sure there's a lesson in there somewhere," My attention faded back to the job at hand.

The design event was a rip-off for us, but it was admittedly partly our fault. On my side of things we had aerodynamic test data for the wings themselves and we had computer analysis, but we didn't have some of the numbers we should have, like full car drag coefficient and downforce dynamics. I was determined to change that before California. The competition might have been partly luck, but the more prepared we could be, the less we would have to hope that luck would show up. Sometimes there were teams that won only because they had that extra bit of luck. They were always good teams, but sometimes the first place team was not the best and everyone could see it. Top five, but not top one. Other times, the best team would win with no luck needed. Those were the great teams, the well prepared teams. The teams that took ahold of their own fate and left the rest of us to hope we could luck our way into second place.

Testing was not finished before Detroit. Arguably, testing could never be finished. We could test for a solid year, only to have to go back and remake parts based on our testing and end up with a completely different car that could be better with just a little more testing. Still, we had time between the events for just a little more. My part of the additional testing was aero. We knew it made the car faster, and we knew why, but we still had work to do. I set up a series of tests to find the actual car downforce and drag at different speeds. We had sensors on the shocks that measured how far the car was pushed down, and since we could calculate the force required to push the car down that much, we knew our downforce. There are better ways to measure downforce, but for a low budget team, shock potentiometers fit the bill pretty well.

I waited until it was late in the evening when everyone had gone for the day. I drove around our local industrial complex, accelerating to just under 100 MPH, then pulling in the clutch and coasting down until the end of the road. I repeated for slower speeds to get the full range, and let the data acquisition computer gather all the necessary information. Going 100 MPH down a city road with curbs and mailboxes was pretty intense, especially since I had a short distance to accelerate and stop. Putting up cones to prevent other cars from entering would have probably been wise, but stuff like that gets the cops called. While technically not legal, we were mostly safe and in the middle of nowhere, so we prepared everything thoroughly and waited until the place was about empty, then did our test blitzkrieg style.

After the test, I collected the raw data from the onboard computer for analysis. I plugged in the appropriate equations and variables into Excel and started plotting graphs to find trends.

The data wasn't neat and tidy, it never is. It looked like a graph drawn by a crack addict in a paint shaker. But the trends were there; and the information we had was plenty to make pretty graphs and show the judges that the wings were advantageous and we knew how, why, and how much they worked.

I overlaid the test data onto our previous simulation graphs. They were close, but could be closer. I started thinking about how the wings affected the car. The downforce lets the tires grip more, but that is only a benefit in the corner, where the outside tire has more load than the inside. So I took out my Race Car Vehicle Dynamics book and added functions to account for lateral load transfer. Then I thought about the change in the friction coefficient of the tires at different loads, so I added that. The simulation data was getting closer to the test numbers. I thought about how the wings may have a different effect with different tires, so I added a selection for different tires. I thought about the decrease in front downforce at increasing speeds, and I thought about the change in the height of the center of pressure. At five o'clock in the morning, I realized I had spent another all-nighter on the racecar. Simultaneously thankful and disappointed that it would probably be my last race team all-nighter, I saved my spreadsheet, inescapable.xls, and went to sleep.

Three weeks later, we were all at the shop early, trying to stay on schedule. We had spent the previous day doing some pre-packing, but we had lost focus at some point and had spent the better part of the evening cooking hotdogs and playing soccer four square.

After finishing up our last bit of packing, I jumped in the back of Ricky's van, and we were off. Ricky had taken out the rear seat of the minivan and had placed a giant pillow in the back which made a nice bed for an after-driving nap. It was a big pillow, but not the world's largest pillow. That was back at Ricky's house next to his Guinness certificate that read "World's largest pillow."

Ricky was an eccentric dude.

We switched off driving duties every few hours, driving all through the day and night. Our average speed had been slightly higher than the posted speed limit, so when we called the others after breakfast, we were not too surprised that we were ahead of them by three hours.

"Let's take a scenic detour," I suggested, lying down in the back of the minivan.

"Where?" Kyle asked. "We're in the middle of the desert."

"We could follow Route 66 for a while."

Silence.

"I dunno, I'm just bored."

"I could go for a detour," Ricky said from the driver's seat, "I'm tired of the highway and I don't really want to sit in the hotel parking lot for three hours waiting for them."

I sat up and searched for a suitable detour on the map. "Take the exit for Seligman, and then we'll follow 66 around and catch back up with I-40 in Kingman."

"Who in their right mind would live out here?" I asked, looking at the mobile homes placed far apart and miles from anything but the highway and each other. They were all far enough from the interstate so that you couldn't really see any distinguishing features and none of them looked like they had a driveway.

"I bet the land is cheap," Kyle said.

"Well yea, 'cause no one smart enough to read a bus schedule would live out here. I bet going out to dinner involves a shovel and some kind of vermin," I replied.

"Hey, I can build my castle out here. The land would practically be free," Ricky announced.

"Screw that, I'm building my castle in the city," I said. "If I can afford the castle, I can afford the land."

"You can't have a castle, I'm building the castle," Ricky said defiantly.

"Whatever, I've always wanted a castle. I wanted one before you," I said with impertinence.

"How do you know, maybe I wanted one before you."

"Well, I'm older than you and I've wanted one for... how

old are you?"

"Twenty-two."

"Twenty-three years," I said authoritatively, sitting back.

"Well, I get the moat," Ricky shot back.

"You don't get the moat; the moat comes with the castle. You can't have a castle without a moat."

Kyle jumped in "How about whoever gets the castle first gets the moat."

"You're going down," Ricky said without looking back, punctuating the statement with a finger pointed down.

"Whatever bitch, I already got fifty bucks saved up."

Near the end of our detour, we got a relief from the boredom in the form of a mountain road. Mountain roads are just like racetracks, except instead of a foam wall to crash into, there is a cliff. Ricky drove us up a mountain and down the other side in a minivan like it was a purpose built hill climb racecar. With the tires squealing at the edge of a cliff and the smell of brakes burning up, I started to think we'd never make it to the race, much less our castles. We decided to continue our detour away from the interstate in hopes of more exciting mountain passes, but all we found were straight roads with the occasional Highway Patrol car deterring us from setting any minivan land speed records.

Twenty miles from the racetrack, a ladder fell from the back of a truck a few cars ahead of us. Ricky swerved slightly to the left, as far as he could without hitting the car next to us, but we clipped the ladder with the right rear tire at 70 MPH. The tire instantly went flat and Ricky pulled over.

As we were coming to a stop, I said, "Fifteen hundred miles with no problem, and we get a flat twenty minutes from the hotel."

"At least we missed it with the front tire," Ricky said, "we only have one spare."

I got out and went to the back to get the spare tire. I checked under the back of the car, but there was no spare tire.

"Not there?" Kyle asked.

"No, must be back here somewhere." I started lifting floor mats with no luck. Kyle opened one of the side compartments and found the jack.

"Where the hell is the spare tire? Ricky, does this thing have run-flat tires?" I asked.

"I don't think so. Look under the back."

"I checked there."

Ricky walked back with an I'll-do-it-myself attitude. He was more subtle about it than I was, and since he usually did know what he was doing I deferred the tire finding duties to him.

Five seconds later he said, "What the hell?"

Kyle had resorted to reading the manual, and with his finger pointing at the passage announced, "It's in the floor behind the driver's seat."

We looked at each other with surprised confusion. *Why would they put it there?* I thought.

"Well I'll be damned," I said, lifting up the floor mat and the carpet that was below it. Twenty minutes later we were standing in the parking lot of the local tire store wondering what to do for the next 45 minutes while our tire was replaced. I looked at the mountain range to the north, amazed that something that big could exist. Norman Oklahoma is flat, and our tallest building is barely over 10 stories, so seeing mountains in the background of the city was just amazing. It was awesome, in the true sense of the word; I was in awe, like an Amish kid in New York City.

"Phil is forty five minutes behind us, so they should get here when the van is finished," Ricky said, looking at his phone.

"Let's go get some food, I'm starving," Kyle said.

14

California

"When I started racing my father told me, 'Cristiano, nobody has three balls but some people have two very good ones.'" – Cristiano Da Matta

Southern California was sunnier and just generally a happier place to be than Detroit. Most of the teams only went to one competition, but there were a few of us who had gone to both events, including Kansas and Missouri-Rolla, and we all appreciated the California weather.

Tech inspection in California was a breeze; we had already fixed all the problems that we had run into in Detroit. We had also purposefully left two things wrong on the car again for the inspectors to find so that tech would go as smoothly as possible. One of these things was found immediately and a team member was sent back to the paddock to "make" the part needed. The other thing, the padding pieces for the rear roll bar supports, was never found. We decided it was just extra weight and left them off.

Unfortunately, while tech inspection was going smoothly, we had a big problem turn up. While waiting in line for the tilt table, Ricky noticed a crack in the sprocket hub.

The sprocket hub is the aluminum piece that holds the rear sprocket to the differential. It transmits all of the power from the engine, so if it fails, the car won't move. Drivetrain parts get a lot of impact and fatigue, and it is hard to design them for varying shifting forces and students who rev up the engine and jam the car into gear at 8000 RPM. This one was failing in a way that we liked to see things fail: slowly and not catastrophically. Still, the timing was bad.

"That's not good," Kyle said, as the team stood around the car surveying the damage.

"Well, it's only in one spot," I said, fishing for comments, "and we don't know how long it's been there. I mean if it's been there for a month, then maybe it's okay?"

Bobby responded, "I think someone would have noticed it when we were cleaning the car before we left, or at least in Detroit."

I took a deep breath and thought about our options. "Well, we still have the brake tech to do, maybe we should see if that makes it noticeably worse."

"I don't really feel comfortable driving all the events with that," Ricky said.

"So, it's a CNC lathe operation followed by a complicated CNC mill operation. The chances that we could get that done overnight, even in Los Angeles, are slim at best," I said, trying not to sound argumentative.

Ricky suggested, "We could fly Chris back to Oklahoma on the next plane out. He could machine a new one there and fly back in the morning."

"I'm okay with that," Chris replied, "but let's see if there are any schools around here that would let us use their shop."

"Chris has to drive tomorrow," Kyle said, "so we don't need him tired from machining all night unless we have to."

Chris and a few others went to talk with the schools that had campuses within driving distance to see if anyone could help us.

Meanwhile, Kyle, Ricky, Bobby and I took the car to the noise and brake tech area. We passed noise on our first try, like it was never even an issue. Kyle then did the brake tech, and again we passed on the first try. As soon as he pulled off the course, I put our last tech inspection sticker on the car while Bobby and Ricky looked at the sprocket hub.

Ricky announced, "Yep, there it is. Another crack."

"Well that pretty much settles it," Kyle said.

As we got back to the trailer, Wes came over to the car and said, "Chris found a school that is going to help us."

Kyle asked, "Who is it?"

"He said CSUN, I think."

"Oh, Cal State Northridge," Bobby clarified.

"Phil and Brett left with them a few minutes ago, so we should have the new hub in the morning," Wes said with confidence, "Chris is going to stay here so he can get some sleep and be ready to drive tomorrow."

"So, I know that hub lasted through Detroit, but are we making any changes to ensure it won't break again?"

"Yea, Chris changed something," Wes turned back and yelled in Chris's direction, "Hey Chris, what did you change on that hub?"

Chris, walking towards us, said, "We just took out the lightening pattern, it will make it stronger and it makes the machining easier. And it's only like one tenth of a pound heavier." He confidently added, "I'm calling it good."

After we finished tech, we went to the design event. It went smoother than it did in Detroit for me, mostly because I didn't get any questions. Everyone else on the team was confident that they had done well, and we felt sure we would make it into semifinals.

We spent the rest of the morning relaxing. Bobby had bought a pirate ship themed kids pool and brought some water misters from home, so we filled the pool and relaxed under the cool mist.

It was like a vacation, we were just lounging around for hours, getting California tans and soaking our feet in the pool water.

"Hey, they printed it this time!" Ricky exclaimed.

"Printed what?" Kyle asked before noticing the event program in Ricky's hands, "Oh, the info for our car?"

"Yep," Ricky said, showing everyone the page in the program that had a picture of our car and a list of ridiculous entries. We all laughed, remembering all the stupid things we had come up with.

"Ha! At the wheel-lair. That's just funny," I said, "Did we check the program at Detroit?"

Ricky answered, "Yea, we checked. They left everything blank."

Kyle's parents and grandparents showed up with a full size RV. When we got too hot from lounging around in the sun, we just went into the air conditioned motor home and grabbed a cold drink from the fridge. It was like being on a professional race team. We watched other race teams run by, pushing their cars and panicking about this or that, just like we had done in Detroit. It was a nice change.

"Attention all teams, attention all teams," A voice came over the loudspeaker, "The following teams have been selected for design semi-finals, and need to be in the design building in thirty minutes."

We all sat quietly, waiting to hear our name.

"Car number seventy-nine, The University of Kansas. Car number one, Texas A&M. Car number four, the University of Washington. Car number eight, the University of Oklahoma..."

"Alright, let's get the car over there," I said with no hint of excitement in my voice.

The announcer continued listing off schools as everyone from our team slowly got up from their relaxed lounging positions around the pool and started to put on their shoes and shirts. Wes was ready to go, so I handed him the poster boards and said,

"Wes, can you take these and go claim us a good spot please." I grabbed my computer and followed Wes to the design building.

The previous year was the first year Oklahoma had made design semi-finals since the early 90's. We were so surprised; we had been preparing so long to make it to semi-finals that we had no idea what to do once we got there. The preliminary design event is a chance to tell the judges about the basics of the car and try to convince them that you know what you're doing and that you have more to tell them in the semi-finals event. Semi-finals is where they ask questions to see if you really did think through every little thing and, more importantly, whether or not you actually understand the system. Fortunately for me, there was very little focus on aerodynamics. Unfortunately for Trey and Kyle, there was a lot of focus on suspension, tire and chassis. We had enjoyed the same experience the previous year when I was in charge of the engine system. Almost all of the judges were interested in the suspension or chassis, and one out of every five judges had some basic engine questions. We didn't get any aero question that year even though James had poster boards and design data for the wings that never made it on the car. So I wasn't expecting much, but I had prepared anyway.

We were all prepared. Everyone on the team was upset about the design event in Detroit, but we had all taken it as motivation to make damn sure we knew as much as possible about our systems and our car.

As soon as the event started, Claude came up to our car and started asking questions. Trey and Kyle stood with their hands behind their backs, occasionally pointing to a graph on one of the posters. I tried to listen in but I couldn't hear much; I was trying to stay far enough back to allow the judges free reign around the car. Occasionally I heard something from Kyle that reminded me how I wouldn't know what they were talking about anyway.

"... when the car transitions from yaw to pitch the mu of

the unloaded rear tire increases, according to the Pajacka model, because the kinematics combined with the compliance of the outboard assembly allows the slip angle…"

I stood patiently in the back, taking an occasional drink from my water bottle and waiting for someone to ask, "Who is your aerodynamics guy?" An hour and two water bottles later, I got my chance.

"Who designed the aerodynamics?" Asked a short fat man in a grey hat that read "Ford."

I raised my hand, even though I hadn't actually "designed" the aerodynamics. Since the guy who had done most of the design had graduated a year earlier and was a thousand miles away, I decided I would represent him.

"Do these things really help?" he asked, pointing to the wings.

"Yes, they give us about a hundred and fifty pounds of downforce at sixty miles an hour."

"Yes, but do they make the car faster?"

"We see a decrease of between one and two seconds on a typical FSAE style course."

His eyes got really wide, still looking at the wings.

"Wow," he said nodding his head, before looking at me and asking, "Who is your engine guy?" I pointed towards Chris.

"Thanks," he said and walked away.

"That's it?" I said quietly to Bobby, "That's my aero question?"

"Be careful what you wish for," he responded, pointing to Claude, still grilling Kyle and Trey with questions.

"Man, he either really likes us or really hates us," I said.

Bobby was watching Mike O'Neil, the head design judge, sitting in Michigan State's car fully strapped in with a helmet on. O'Neil was banging his helmet off the roll hoop, presumably checking for appropriate head movement and padding.

With a worried tone, Bobby said, "Oh shit." I was confused for a moment, and then suddenly realized why Bobby was worried.

Ricky apparently suddenly remembered as well. "We need to get that roll bar padding on," He said quickly, but in his usual calm voice. The design judges would frequently pull tech stickers if they found something on the car that wouldn't pass technical inspection. Even though we had already passed, if O'Neil didn't like our unpadded roll hoop supports, we would have to go through tech all over. We had time, but would rather spend it around the pool than stressing over another round of inspections.

Ricky leaned over to Wes and asked him to go get the roll bar padding from the trailer. As Wes was walking off, Ricky said, "Hurry," and did a running in place mime. Wes started running just as O'Neil walked up.

"Okay guys, I'm going to get in your car and judge your ergonomics, so do what you need to do so I can get in."

Ricky reached down, pulled off our quick release steering wheel and said, "Go right ahead, sir."

I grabbed a couple zip ties and handed one to Bobby. O'Neil was pressing the pedals and turning the steering wheel, checking the feel from the driver's seat. I wasn't worried about scoring poorly on this part; our car was the most comfortable car I'd ever sat in. I wanted to make a lounge chair using our seat mold. You could fall asleep in there.

Wes ran up and handed the two padding pieces to Ricky, who passed them off to Bobby and me. O'Neil had his helmet on, so we snuck up behind him, each put our padding on the rear roll bar support, and zip tied it on. Not two seconds later, O'Neil banged his helmet side to side, testing the padding.

O'Neil got out and moved on to the next car. Bobby and I cut the zip ties and wrapped black tape around the padding to make it sturdier.

"Wow that was close," I said quietly to Bobby.

I barely finished the sentence before I noticed Kyle, who was still talking to Claude, pointing at me. Claude started walking in my direction.

"Hi Mott, it's good to finally meet you,"

Claude's accent was not very heavy, but there were words he just never said right, and my name was one of them.

I nodded and said, "You too."

It was all I could think of.

He paused for a moment and said, "Tell me about your irowdynomics. How did your simulations compare with your testing?"

"Well, our simulations showed our downforce to be..." I walked towards our poster that had all the aerodynamics graphs and started explaining our simulation methods, including the spreadsheet I had spent the better part of a sleepless night working on. He patiently waited for me to finish explaining before asking another a question, a courtesy not common among design judges. He also asked direct questions instead of letting me guess what he was looking for.

"How did you choose your irowdynamic goals with respect to downforce and drag?"

We went back and forth for about three minutes before he said, "It was good talking with you Mott. I am very impressed. I need to spend some time with the rest of the teams, but we will talk again later."

"In design finals, I hope," I said with a smile as he was walking away. He shrugged his shoulders and continued walking.

Two other judges had grabbed Kyle and Trey while I was talking to Claude. Chris was busy explaining his drivetrain design to the owner of the company that made a few of the drivetrain components we were using.

I walked back to where Bobby was standing.

"Well that went okay."

"Yea?" he asked.

"I think so. Better than it could have gone, for sure."

At the end of the day, we packed up and headed back to the

hotel. Bobby called Phil to see how the machining was going. They were just finishing up the coding for the lathe operation, which probably meant an all-nighter.

The next morning, just as we were all waking up and getting ready, Phil and Brett showed up with our new hub piece.

Phil looked like he could stay up all day, but Bobby insisted that they stay at the hotel and get some sleep for a few hours.

"If you see Pablo and Rachel there, tell them thank you again," Phil said.

"Who are Pablo and Rachel?" I asked.

"They are the two CSUN people who stayed up all night machining with us," Brett said, looking exhausted, "Rachel is the girl with pink hair and Pablo is... the guy named Pablo."

We arrived at the track as soon as the gates opened. Chris took apart the drivetrain, replaced the hub, and put it all back together before the events even started.

"Do we need to take this out to the practice area and test it out?" I asked.

"If we just do the skidpad and acceleration runs early, we can use those events as a test and maybe still have time to fix something if we need to," Chris said.

"If it breaks, it's not like we can fix it," Kyle said.

"We can always just weld the ever loving shit out of it," I said, "It'll ruin the heat treat, but it's better than nothing."

Ricky confidently said, "I don't think we'll have a problem with it, it's stronger than the last one, and that one lasted way longer than we're going to drive it this week."

Ricky, Chris, Bobby, and Brett took the car out to the event area. They waited for a few cars to run the events and sweep the dirt off the track before they went out. Our acceleration time was a little better than it was in Detroit, not bad but not great. Bobby got a time that put us in 13th place. Our skidpad time was also like Detroit, Chris put down a third place time. They checked the sprocket hub after each run and every time it looked perfect.

They brought the car back and we all stood around it for a moment before realizing that there was nothing to do. So we all gathered around the pool again, drinking Gatorade and soaking our feet in the water. Brett and Phil showed up just before lunch and joined us. Michelle and Gloria had again set out lunch for us, and we took turns making sandwiches before gathering back around the pool to eat.

"You all need to make sure and go thank CSUN for their help," Gloria said.

"We did," Phil said, "we told them that next year we would have a sticker on the car that said 'It's all about Pablo and Rachel'."

After lunch, Kyle and I went out to walk the course before the autocross event. There were a couple areas where we both agreed there might be trouble: one very tight corner where more than a few teams would take out the apex cone, and one increasing radius turn where there was a different asphalt surface halfway through. Different surfaces would give different grip and might upset the car. The effect was very noticeable from concrete to asphalt, but it could still be a problem with two different types of asphalt.

We went back to our paddock and waited for several other cars to go out before we pushed the car to the track. I was much more relaxed for the autocross event than I had been in Detroit. I'm sure the lounging around the pool and the RV didn't hurt.

After watching a few cars go around the track, I got in and started the car to get it up to temperature. I sat with my eyes closed, going over the track in my head. This course was a lot tighter than the Detroit course, which meant our wings wouldn't be as much of an advantage.

Deep breath, immediate left, small sweeping right, slalom...

After a minute or so, I said, "Let's do this."

"Now? Do you want to wait for a few more cars to go and

clean off the track?" Kyle asked.

"No, I'm relaxed and ready. The track is pretty much clean and I don't think it's going to get any hotter out."

Brett and Bobby pushed me up to the line. As the cars in front of me went out, I eased the car forward, stopped, and continued to go over the track in my head. I watched the two cars in front of me to see if they were having obvious trouble with any corners. As soon as I got the green flag I was ready to go.

I took a deep breath and accelerated away. Immediate left, sweeping right, slalom, tight right with an increasing radius and a transition onto different asphalt. The change in grip was noticeable for the different surface, but not enough to upset the car too much. I finished the course without finding any areas for great improvement. I went back out immediately and took a couple of the corners a little bit faster. My improved second lap was over a second faster than my first.

I pulled the car to where Brett and Bobby were standing with the push-bar. I shut off the car and asked, "Where does that put us?"

"Fastest time of the day," Bobby said matter-of-factly.

Our on track performance continued to parallel our Detroit performance. We stayed in first as we waited for all the schools to go out and finish their first run. During their second driver's run, Michigan State took FTD from us, and then Texas A&M took it from them. Kyle went out and almost took first place back. He was four hundredths of a second behind A&M, literally the blink of an eye.

15

Panoramic

"The flowers of victory belong in many vases." – Michael
Schumacher

Kyle and I went to the driver's meeting for the endurance event. Every event had a driver's meeting, and we had been to them in Detroit and the previous year so we knew what to expect. They all said basically the same thing: the yellow flag means caution, the red flag means stop, and the white flag means there is a slow car ahead. They always made extra clarification for the last two.

"The white flag does NOT mean one lap to go. This is not circle-track racing. If you see the white flag, it means there is a slow car ahead. Do NOT pull into the finish after you see the white flag unless you have been given the one-lap-to-go signal."

"The red flag means slow down safely, pull to the side of the track, and stop. It does not mean slam on the brakes and skid into a corner worker."

All of the students were thinking the same thing, and one of them brought it up.

"What happens to our lap times if the red flag comes out?"

The official calmed everyone's worries, "You'll get to restart from the beginning of that lap and the red flag lap won't count. Honestly, we never throw the red flag in the endurance race. I can't remember the last time we did; it must have been ten years ago. I wouldn't even worry about it."

After the meeting, Kyle and I headed back to the trailer. Just as we got there, the announcer came over the loudspeaker and said, "Attention teams, attention teams, the following teams have been selected for design finals."

The man paused. I looked around at everyone listening intently and quietly staring in slightly different directions.

"Car forty-two, Brigham Young University... Car seventy-eight, Michigan State University, Car eight University of Oklah—"

We all yelled out at the same time, before he even finished our school name.

"Boomer!" Gloria yelled.

We all responded with a very loud, "Sooner!"

I think we all expected to make it into finals, but we were still amazed. The goal was to win the competition, but the design event was a competition in and of itself. It was the meeting place of the very best teams, and we just got an invitation.

Michelle, Kyle's mom, walked up to me and asked, "What size pants do you wear?"

"Uhhh, what?" I asked, thinking I misheard.

"We're getting khaki pants for the team for design finals so that you all match and look nice."

"Ooooh... thirty-two thirty-two. Or thirty-four thirty-two, it depends on the pants."

She laughed, "Kyle is the same way so I'll just get one of each and you guys can decide."

"Sounds good," I replied. As she walked away I said, "Thank you."

We spent the rest of the afternoon sitting around the pool, with smiles plastered on all of our faces. Dr. Gollahalli, the Director of the School of Aerospace and Mechanical Engineering at OU, had flown in to support the team, and was enjoying the California weather with us around the pool.

Brett had added up some of the scores, and it looked like we had a very good chance at winning the competition. After all our work and effort, four years of sleepless nights and impossible obstacles, we were only one day away from the finish line.

As late afternoon approached, there was an announcement that the teams needed to gather on the track for a picture. Every year, all the teams gather together with their cars for a panoramic picture. It is exciting to see all the competitors in one place, around a hundred different cars all very different and the thousands of students that built them. It's easy to get caught up in the details of the competition and not realize the scope of it. The design finalists always got up front in the center, and we were finally one of those teams. Kyle didn't want to rely on the good graces of the other teams to save our place.

"Get the car, let's go," he said, waving the car along, "We need to get there before someone else takes our spot. Hurry up."

Wes and Brett started pushing the car as Bobby slipped down into the driver's seat. Kyle went ahead, presumably to save our spot. All of our events were held on the infield track, but the picture was being taken on the oval, right at the start/finish line. To get to the start/finish line, the cars had to be pushed through the infield grass. Bobby got out of the car and helped them push it along.

"You think they'll let us do donuts in the infield if we win?" Ricky asked.

"I doubt it," I said with a smile. I was thinking the same thing, and I'm sure everyone else was, too.

After we had our car in position up at the front, I noticed that one of the teams was dragging their car across the grass with one wheel missing. They had met with some bad luck earlier that day when their suspension failed and a wheel broke off. With two people pushing and two people holding up the car by what was left of the right front suspension, they slowly pulled it across the infield. When they got out in the open where everyone could see, a cheer started. Slowly at first, then everyone looked up to see the team diligently pulling their wrecked car across the grass and soon everyone was cheering. One of the guys holding up the front of the car pumped his fist in the air in triumph and then went back to pulling. Even after a catastrophic failure, there was still a little success for this team in refusing to give up and go home.

The panoramic picture was the only time that all the teams were together in the same place. Hundreds of students, all exhausted and tired, gathered around their respective cars. I thought about how we had all taken completely different paths to end up in the same place. Except the team with the broken car, they were still in the grass.

When they finally made it to the track and we were all lined up, the camera man snapped a couple pictures, announced that he was taking orders for printed pictures, and thanked us for our patience.

I half-jokingly said, "We should run around the oval. The whole two miles."

Wes shrugged his shoulders and said, "Okay."

I assumed that running the track would be a bad idea, seeing as how they had told us not to go anywhere but a few specifically designated areas. But Wes was usually a pretty good judge of right and wrong, so I figured if he was up for it, then it was probably okay. Or at least not terribly not okay. We lined up at the start/finish line and started running. I thought about how we should have used a stopwatch to keep track of our time so we could compare it to the oval track car times. You know, their 46 seconds to our 20 minutes. After a short distance, I was out of breath.

"Hold on, lemme... catch my... breath," I said as I started to walk.

"We just started," Wes said with a laugh.

I said, "alright," with a reluctant tone and started running again. After another minute, I said through my breath, "Man I'm outta shape. This is a long freaking track."

I made it another few hundred feet before I stopped again.

"I ca... I can't...nogonna make it," I leaned over to catch my breath. Wes stopped, clearly ready to continue the run. I could still see the starting line. I looked up at the wall to a giant painted sign that said "Turn 1."

"I didn't even... make it past turn one," I said, leaning over and trying to catch my breath. "I'm glad they let us use cars in this race." I looked forward and thought really hard about finishing. I thought about how I'd have to make it to turn two, then three, and then four, then a short stint to the finish line. It was a long run, but I really hated the idea of giving up, even when I decide to do something stupid. Somewhere between crazy determination and reluctance stemming from assured failure, a truck started towards us. *Oh thank god*, I thought. *I get to quit without quitting.*

"You guys can't be on the track, you need to go back," the driver of the truck said.

We agreed and he drove off.

"Man, I was just about to finish the whole thing in ten minutes," I said with a smile.

"Uh huh," Wes replied.

When we got back to the paddock, Bobby asked "Did you make it?"

"We made it to turn one, does that count?" I said.

"The track security stopped us," Wes added.

"Not that I would have made it much further. I'll have to practice running and come back next year and try again."

"After you've graduated?" Wes asked.

"I'll come back as an alum."

"Even if you live in Detroit?"

"Wes…" I said smiling and shaking my head, "Just let me have my fantasy of someday being in shape."

I wasn't really in terrible shape; the race weekend wears a person out. Team members sometimes talked about being in a fuzzy haze after the competitions. Partly it was the realization that you've reached the end of that year's all-consuming obsession, but mostly it was the twenty hours a day of exhausting stress endured for a solid week. Our week so far was relatively laid back compared to Detroit, but it was still draining. We had been sleeping very little and running around in the sun pushing a car a mile at a time, surviving off energy drinks and stubborn relentlessness. Even while relaxing by the pool we were baking in the sun, our energy draining away. In retrospect, it was a good thing we didn't get to finish the run, because the next day was the endurance event which really would be a workout.

We finished packing up our stuff and headed back to the hotel. The next morning was the start of a big day and we needed to get whatever sleep we could.

16

Dancing on the Edge

"And so you touch this limit, something happens and you suddenly can go a little bit further. With your mind power, your determination, your instinct, and the experience as well, you can fly very high." – Ayrton Senna

I started the car to get it up to temperature. Chris had his laptop plugged into the car's computer so he could tell when the engine was warm enough.

"It is so freaking hot in this driving suit," I said with my head leaned back and my eyes closed. Our driving suit was a thick black fireproof outfit that would have been much better suited for use in the Iditarod race. It was not enjoyable when the temperature exceeded 80, although I'm sure it would have been greatly appreciated if the temperature exceeded 450. "You think they'll let me drive in my underwear?"

I was just talking for the sake of keeping my mind occupied, trying to stave off the nervous feeling.

Trey shrugged his shoulders, not amused.

"Hand me that water," I pointed to a half empty water bottle in our cart. I unzipped the top part of my driving suit.

He handed me the water and said, "They told us not to do that last year; it feels good at first but is uncomfortable later on."

Already wet from the sweat of intense heat and nervous anticipation, I didn't see any loss of comfort arising from a brief shower. I poured the water down the front of my shirt and basked in the chilled refreshing feeling. I left the top unzipped so the water could slowly evaporate and cool down my shirt even more.

"Ahhh, latent heat of vaporization," I said quietly.

"What?" Trey asked.

"Never mind. When are they gonna start this thing?"

Like that was his cue, a man with a radio in one hand and a clipboard tucked under his arm yelled, "Listen up everyone! We are going to start in about five minutes!" He walked down the line of cars holding out five fingers and repeating himself.

I took a deep breath.

I sat quietly for a moment, trying to stay calm. Kyle had stopped talking to me; he must have seen how nervous I was. My heartbeat could pump dry a swimming pool. After four years of building these cars, four years of stubborn persistence and bad grades, deleterious arguments and late night engineering, we were just 20 laps away from the end; 20 laps away from a first place trophy. I hadn't really seen the progress that we had made over the previous few years. When you're so focused on the details, it's difficult to step back and get some perspective. But it was all clear now.

I thought back to the first competition I had gone to, looking at all the design finalists and the teams that were speeding around the track faster than everyone. They seemed so far away, a level that was unreachable by a relatively unknown state school in Oklahoma. But we had done it; we had reached that top level. One of the fastest cars on track and one of the top designed cars. Now all I had to do was drive fast and not screw it all up.

I took a deep breath, closed my eyes and went through the course in my head. Out to the left, accelerate in first gear to an immediate right, coast through the middle of turn one and accelerate after the second apex. I went through the whole course over and over. I couldn't keep my focus for more than one lap, but after every distraction I would bring myself back to the track. They say practice makes perfect, but only perfect practice makes perfect and the only way to practice perfectly is to do it in your head. Unfortunately, there were always parts of the track where you don't know what to do just by looking at it. You just have to drive into it with faith and experience and figure it out on the way. It's always a battle between that little devil on your right shoulder telling you to go faster than hell, and your mom on your left shoulder reminding you that you don't have health insurance.

There were two parts of the endurance that were like this. One of these parts was a slalom with an option. Usually slaloms have a pointer cone at the beginning that is laid on its side telling you which direction to go for the first cone, but sometimes they just leave it up to the driver. It doesn't matter which side you start on as long as you oscillate the rest of the slalom cones. Usually it is obvious which side is the fastest, but sometimes you have to figure out how fast you will be going out of the preceding corner. In other words, you have to call an audible while driving into the slalom at full speed.

Blue forty two! Blue forty two! Go to the left! Set hut!

The other part of the endurance that I wasn't sure about was a section that was basically two lane changes. There was a straight away, then a wall of cones where you had to quickly switch to the left side, then another wall where you would switch back to the right. At the beginning of this was a slalom preceded by a sharp turn. So you would exit the turn, do the slalom, and then the lane changes. Each of these maneuvers was faster than the last and I had a feeling that I could do the whole thing with the accelerator pedal planted to the floor. The only problem with my plan was that by the time I got to the last lane change I would be going about 70 miles per hour. Any off course excursion would result in the car plowing sideways through about forty cones. I decided that I would start off easy and maybe try it after a few laps if I felt comfortable.

Chris stepped into my field of vision and waved his hand in front of his throat, telling me to shut off the car. He closed his laptop and unplugged it from the car.

"Two minutes!" the man with the clipboard announced.

My heart started beating fast again.

"Alright Superfast, just like we practiced. Go fast and don't hit any cones," Trey said.

I felt like I was going to throw up. The idea didn't bother me so much, I just wanted to drive. I brought my attention back to the track. Out to the left, accelerate in first gear to an immediate right, coast through turn one and accelerate after the second apex. Enter turn two on the outside and take it wide, carrying as much speed as possible into the straight. Hard on the brakes and trail brake into the corner. *I shouldn't have drunk so much water.* I thought about how hot it was. *Back to the track.* Trail brake into the corner and enter the slalom on the inside. Texas A&M's driver started their car and pulled to the starting line. I started the car and put it into first gear. I looked around to make sure everything was out of the way, and I pulled up to where A&M had been. They got the green flag and took off onto the track. The man with the green flag waved me forward. I drove forward slowly and stopped. He was watching A&M on the track when he waved the green flag for me. I took a deep breath.

I let out the clutch and jolted onto the track. Accelerate to an immediate right, coast through turn one. Coasting through turn one became sliding sideways through turn one. The tires were still very cold and low on grip.

"Okay Matt, a little less enthusiasm and a little more feel," I said to myself.

I entered the outside of turn two and carried as much speed as I could, feeling the car sliding slightly sideways. I was right on the edge. I accelerated out of the corner and hit the brakes a bit too hard entering the following corner. With an open wheeled car like ours, it was easy to tell when the tires were locked up, because those two big black round things right in front of you would stop spinning. That was my cue to ease off the brakes a bit. I took the next corner pretty well and entered the slalom. I accelerated through the slalom, easing on the accelerator more and more until I was wide open at the exit. I eased off the pedal slightly for the first lane change, moved over and put my foot back down. I eased off for the second lane change and did the same.

"I could probably do that wide open," I said to myself.

I took the hairpin perfectly, coming so close to the apex that I could feel the tire driving over the base of the cone. The butter zone. I continued through the rest of the track, going over the course in my head, feeling the tires get more grip as they heated up. The sick feeling I had was long gone, completely forgotten, along with all the nervousness and all the people standing around watching. It was just me, the car, and four hundred cones.

I came around the last sweeper before the option slalom. I was staring at the first cone all the way through the corner, watching the sweeper cones in my peripheral vision.

"Left?" I said, "no, right....right?"

As soon as I got to the point where I had to make a decision, it was clear. I went left. I thought about how some day I would come up empty right before the slalom and plow straight through the first cone.

"Back to the race, Matt. Focus."

I went through the start/finish lights, watching the flag man to make sure I didn't get any flags. I went into the first corner. This time the tires were warm and I had a much better result. Going into the lane change maneuvers, I pushed the pedal a little harder.

"Yea... I can take that wide open," I said to myself, hoping the comment would instill the confidence.

There were about four cars on the course by that time, each spread out pretty evenly, but I could see myself getting closer to the car in front of me. Every time I went through the one big corner, the sweeper, I gained several car lengths on them. God bless aerodynamics. I focused on improving a couple corners, feeling out the car and thinking of how to take them faster the next time.

"First gear exit there," I said, making a note to myself, "Closer to the inside on turn six."

Driving around the corner right before the slalom and lane changes, I said, "Let's do it!" committing myself and cementing the determination to see it through, pedal to the floor, all the way.

I put my foot down a bit harder and drove through the slalom as fast as it would go. The devil on my shoulder was laughing feverishly as the tires slid at the edge of their traction limit. I was on the base of every single slalom cone, just enough to not tip them over. It was Butter City and I was the mayor. Foot to the floor I went into the first lane change. The car slid sideways a bit coming onto the straight and I was going faster than hell. I thought about how I was going too fast, but I had committed myself. I had this lingering feeling in the back of my head that there was no way the car would react to the input I was about to give it, but I did it anyway. I jerked the wheel to the right, then back to straighten out the car. The whole car slid sideways down the track, all four tires were out of control. I was going 70 miles per hour and my left tires were inches from plowing over one cone after another.

"Shit! Shit! Fuck! Shit! Fuck!"

I couldn't counter steer into the turn or I would drive right through the cones. I held onto the steering wheel like a million dollar lottery ticket and leaned to the right as though I was in danger of being hit by the cones myself. The tires regained grip, and I quickly moved back to the center of the track.

I took a quick deep breath, planted my foot into the brakes for the next corner, and brought my focus back to the race.

"Late apex, downshift, first gear exit here."

I looked down at the dash to check the warning lights when I saw that the temperature light was on. *Lap three and it's already two hundred thirty five degrees? That's okay, we can do the race at two thirty five, no problem.* I decided to ease up just a bit in some areas. There were a couple of corners where I could let the engine relax and not give up much time. I would lose a few tenths of a second each lap, but I had to make sure the car wouldn't get too hot.

I drove, lap after lap, focusing on driving and balancing speed with minimal engine load. The devil on one shoulder had made his way down to my right foot and was jabbing his pitchfork at it, trying to make me go faster, while on my left shoulder, a Honda engineer was telling me that the engine was never designed for that kind of abuse. Even taking it easy, I was able to pass Texas A&M. Their faster driver was only barely faster than Kyle in the autocross, but I was much faster than their slower driver. Between two drivers each and 20 laps, I knew we would have the best endurance time. After passing A&M, I managed to pass two more cars. We were well on our way to first place. We were winning and I knew it.

In four years of building cars, spending dozens of hours a week, sacrificing grades, and sleep, and that college experience we heard so much about, we had transformed our team from a mediocre 50th place group of college kids into a first place team. We were ready to go on stage and get the big shiny trophy, and then get the big shiny interviews to the big shiny jobs.

"Just a few more laps," I said quietly, "Just a few more."

But it wasn't over yet; I tried to stay calm and focused while still trying to improve every lap.

I came out of a corner and hit the up-shift button to shift into third, but nothing happened. The engine started banging off the rev limiter and I hit the up-shift button again. Still nothing.

"Shit! Come on! Come on, give me a freaking break here!" I yelled. I hit the up-shift and down-shift buttons several times, but nothing. The shifter was broken and I was stuck in second gear. The high speed sections were third and fourth gear territory, and spinning the engine at redline would catastrophically overheat it in no time. I started to pull the clutch in and coast in every straight away. I was losing huge amounts of time, but I couldn't let the engine spin at fifteen thousand RPMs for that long.

"Come on baby, you and me, you and me. We're gonna make it! You and me!" I could feel the acceleration that the car should have had in the straight-aways, I could feel how fast I should have been going, but it wasn't there. It was a slow coast down. I was stuck in the fastest racecar on the track and turning the slowest lap times.

I could hear the engine getting louder, the internal components slapping around and grinding from the oil being too hot. It must have been 250 degrees at least. I thought about how hard it was going to be to restart at that temperature. I was sure the engine would seize as soon as it stopped. I went into another straight, revving the engine to fifteen thousand and pulling in the clutch. Massive acceleration followed by a painfully slow stroll with a wide open track in front of me. I could feel the seat getting hotter from the excess heat coming off of the engine. "COME ON! WE'RE GONNA MAKE IT! COME OOOON!" I yelled at the top of my lungs, rocking back and forth, trying to get that extra little bit of speed. "GOOOOOOOOOO!!!" I started to really worry that the car wasn't going to make it. We weren't even half-way through and it had been overheating since lap three. How was it going to make it ten more laps with Kyle stuck in second gear? Was it even going to restart? "Come on. Just a few more laps. Just a few more. That's not too much, is it? That's not too much." My frustration had turned to anger and then to disappointment. I was going through the stages of grief and I hadn't even gotten out of the car yet.

I put my focus back on the race. Enter wide and carry speed through the corner, accelerate to redline and pull in the clutch. Wait. Wait. Wait. Make a sandwich, read a book. "Fucking GO!!!" yell profanities, hard on the brakes, but not that hard because you're only going 30 miles per hour by now. I passed the flag station and the man behind the barrier gave me the one lap to go signal. I took my last lap very slow, trying to let the car cool off as much as possible for Kyle. I thought about how I was going to explain the situation to him, although he had probably figured out most of it already.

Coming onto the second to last straight, I pulled in the clutch and coasted all the way to the driver change area. I stopped the car, shut off the engine, took off the steering wheel and started yelling.

"It's stuck in second gear! It has been overheating for seven laps! You have to coast in the straight-aways for fucking ever!" Kyle was very laid back as he was getting into the car and strapping in. I think he was trying to remain calm and focused, but I thought we were well beyond good driving and firmly in the land of please-God-just-let-us-finish. "Did you hear me?!" I asked. He nodded. Chris put his hand on my shoulder and pulled me back. I stepped out of the way.

I took my helmet and gloves off and said with panic in my voice, "It's not going to start. There's no way it's going to start."

We stood quietly waiting for Kyle to tighten his seat belts and get ready. I could see the heat coming off the engine, wavering in the air just above. When he was finished, the inspector checked his seatbelt and restraints and gave him the signal to go ahead. He pulled in the clutch and pressed the start button. It cranked over a couple times and fired up.

"Holy shit," I said, standing in shock, "It fucking started," I looked at Chris, "I don't believe it."

Kyle got the green flag and went out on the track. He took the first few corners perfectly. It was good that we switched; he still had the motivation to do well, and I had pretty much lost my focus and emotional control.

Going into the straightaway, he revved it up to redline, pulled in the clutch and coasted. I was feeling anxious just watching him. When he came back by us, I could hear a knocking noise in the engine. Most likely we had seized a connecting rod bearing. Of course by that point, we had probably seized a few connecting rod bearings and a couple crankshaft bearings. It was getting loud because one of the bearings was completely destroyed and the connecting rod was just banging around in the engine with the crankshaft only loosely telling it where to go. There was almost no chance that the engine would survive for nine more laps. Kyle continued to drive the corners as best he could, and coasted in the straight-aways patiently waiting for the corners. I knew that nothing short of a complete engine failure would stop him from trying his best to finish and finish well. He went around the hairpin on the opposite side of the course from us. I could hear the engine quickly spin up and just as it was starting to get to redline, a huge ball of fire exploded from the engine. The car spun out and plowed through a dozen cones before coming to a rest on the edge of the track.

I stood there in shock, watching all the course workers wave their red flags and all the other cars come to a stop. The safety workers sprayed our car with fire extinguishers and the flames died out as quickly as they had started. Kyle got out of the car and walked away.

All the cars were shut off and everything had gone silent. I dropped my helmet and stood in stunned disbelief.

And that was it; the engine was dead. It was over. Our chances at winning were completely gone. All the hours we had worked, all the sleep we had missed and all the classes we had skipped, every moment spent trying to get the car done one time, and trying to pass tech. It was all over. The culmination of all our years on the team was smoldering on the side of the track, covered in oil and fire extinguisher powder.

The track workers pushed our car out of the way and carefully cleaned up the mess we had made, picking up the cones that had been scattered around the otherwise neatly laid track. I slowly walked over to the pit cart, opened a bottle of water and sat down in the shade.

Auto racing is a challenge of dancing on the edge. We danced too close, and we fell off. Some people would say that we did our best and we learned a lot and that's all that matters. Or that it doesn't really matter because it's just a competition, and only one that a few people know about and even fewer care about. But it did matter. In those few days it mattered more than anything. What got us there was the passion we brought, and the drive we had to do the best that we could do, and then do a little more. Since the moment when each of us on the team, as individuals in our own time, looked at Wisconsin and Penn State and Texas A&M and said "I want to be there. I want to be a part of the team that builds the best car. I want to win." we had all given up so much to be there and had spent so much of ourselves to win it all. It was in that seemingly perpetual panic to get a job in racing and that obsessive drive to win that we had all met, become the best of friends, and grown together into the professionals, friends and engineers that we had become. We had lived for the past few years with more passion than we could have doing almost anything else. At no time during the year did we say "It's all just for fun, it doesn't really matter, all we can do is try our best." The teams who say that finish 50th. The path we had taken demanded an obsession with being the best. We danced close to the edge, the edge of our budget, the edge of our time, the edge of our strengths and our weaknesses. Every time we fell off, we learned something about the car, about engineering, about teamwork, and about ourselves. Not because we fell, but because we made the choice to take responsibility for it. Luck sometimes really is the difference between a big trophy and a long quiet trip home, but, as people, we are never defined by that luck and always defined by what we choose to do with the opportunities that it creates.

I heard engines starting and looked up to see the cars that

were still on track parading back to the start. The cones were back in place and our car had been pushed out of the way behind a small building on the far side of the track. At the south side of the track, I saw Claude standing with his arms crossed, looking in my direction. Looking at me.

"Design finals," I said under my breath.

I jumped to my feet and walked quickly to the edge of the track. There was a man standing in the driver change area looking official, so I waved him over.

"We need our car for design finals," I said before he had made it to me.

"You need what now?" he asked, taking off his radio headset.

I spoke louder, over the noise of the racecars driving by, "We have design finals in an hour; we need our ca... our smoldering wreckage!" I pointed to our smoldering wreckage.

He laughed, "Oh!... well... I guess we're going to have to push it around the back!"

"I can send some people back around this way if you can get one of your guys to meet them!" I said as I pointed to the other side of the infield wall.

He looked back at the far side of the track, then back at me, "Okay!"

I jogged over to Kyle, who was standing with two of the Kansas team members.

Before I said anything he said, "We need to get the car for design finals."

"I know," I said, "They have someone that is going to meet someone from our team around the back way. Radio in to Bobby and tell him to send a couple people back to the far side of the infield wall."

He held one side of the headset to his ear and relayed the message to Bobby.

I jogged back to the pit cart and took off my driving suit. Standing there in the California summer heat, wearing only my t-shirt and boxers in the middle of two hundred engineering students I thought, *we're going to design finals*, and managed to squeeze out a half smile.

17

Finals

"The only ones who remember you when you come second are your wife and your dog." – Damon Hill

Before I even got back to the trailer, I saw that the rear wing was folded forward. At first I thought the team was taking it off to clean the engine, but I noticed the supports were cracked in two.

"What the hell, man?" I asked.

Ricky responded, "Yea, we thought you'd be pissed."

"What happened?"

"After the fire went out, the course workers tried to push the car off track by the rear wing," Ricky said with little concern.

I threw up my arms in disgust before he even finished the sentence.

"Oh, you gotta be kidding me. What the hell?"

Looking at my collapsed wing, I noticed a gaping hole in the engine just behind it and suddenly regained my perspective. I shrugged my shoulders.

"Well it's not the worst thing to happen to the car in the last five minutes," I said, and started walking back into the trailer.

Everyone was already wearing their OU shirts and khaki pants. I pulled my shirt out of my backpack, half balled up and half folded. I put it on and pulled it down to try and make the wrinkles disappear, then I let go and watched them all reappear with mild disappointment. Looking around, I realized we matched better if I left the shirt alone. Most of the team had left their shirts on after design prelims and semis, so everyone's shirt had seen a bit more sweat and wear than would be ideal. We actually didn't look too bad for a stressed out team with dirty shirts, and whose hopes for a first place finish had exploded into a ball of fire only moments earlier. Some of the team members had started to clean the oil and fire extinguisher powder off the engine to make the car more presentable. Some of the senior team members were studying up on their systems for a few more minutes. I couldn't study anymore. I suspected it wouldn't matter anyway; unlike the finals in class, design finals can't be edged out with last minute cramming. I grabbed some tape and started to fix the wings, adding splints wrapped in black duct tape as a temporary and cosmetic fix to the rear wing.

"We should get going; someone needs to pick up the posters and stands," Bobby said authoritatively.

I grabbed the posters and asked Brett to get the stands. We headed towards the design area to pick a good spot. When we got there, the first three stalls were already claimed and setup by the other design finalists. We walked to the back corner and setup our stands and posters. I asked Brett to help me get an extra table from the other side of the building for all our laptops. We had most of our important data on the posters, but it was good to have the laptops on and ready just in case.

I scanned our area to make sure everything was in place, took a deep breath, and waited with my arms behind my back and in my best attempt at open and honest body language.

"How did you decide on this wing configuration?"

"You mean the element spacing or…"

"No, I mean one split wing up front and one in the rear."

Okay, think Matt. What is he really asking, what is he trying to find out? He's trying to find out if we looked at other options, and if we did, had we done any analysis or testing to show that this one was better.

"Well, we first looked at some different ideas that some other teams had done, and some new ideas we kicked around in a design meeting —"

"Like what?"

"We considered one large wing above the driver, but the clearance needed for the driver egress would cause the center of gravity to increase more than this setup."

"How much more?"

This guy is relentless.

"I don't… remember specifically; it was much more than this setup, which increases the CG by just over a quarter inch. We also considered using an undertray to increase downforce, but our computer simulations showed it had a relatively small impact on downforce—"

"Compared to what?"

"Compared to the wings alone, and it also required that we move some components like the radiators higher and the tunnels would have intersected with the suspension members. Also, we would likely have to increase the ground clearance. So we went with two separate wings, one split around the nose in the front to make use of ground effect, and one in the rear just high enough to be in clean air," I said confidently.

"Some of the teams have very large wings, did you do any testing for different sizes?"

"We did some analyses, but it's a tradeoff, we had to take a systems approach…" Judges love hearing this. The racecar is a system with every component affecting every system in some way. Bigger wings would create more downforce, but at the expense of other performance variables. This was the question he was asking, had we taken into consideration the affect of the aerodynamics package on other systems and had we chosen a good balance.

I continued, "...We had to take into consideration the affect of the aero package on the other car systems..." he started subtly nodding, "...and based on some downforce numbers from CFD and vehicle simulations, we felt we were in the Goldilocks Zone."

"The Goldilocks Zone?" he asked.

"Yea, not too big, not too little. Just right."

And it went on, back and forth. He would ask a question, and I would try to figure out what he was really asking. The design event can be crudely summarized in two questions. The first is "Does it make the car better, and can you prove it?" When I couldn't figure out exactly what he was looking for, I took the answer in this direction. The second is "Do you *understand why* it makes the car better?" This was less fun for me, because I was not an aerodynamics person. I didn't thoroughly understand the underlying science of aerodynamics, and I only partly knew how it related to racecar performance.

He started asking the second question, and I started to get worried.

"What would happen if you reduced the gap between the elements?"

Well, let's see, I know that James had done some computer simulation on the gap, so the one we chose was the best for our situation. That means changing it would either decrease downforce or increase drag.

I had to stop staring blankly at him and respond.

"It would reduce the efficiency of the wing."

"Yes but how?"

Our wings were like snow shovels, they scooped the air up and threw it upwards; every action has an equal and opposite reaction, so the air going up caused the car to go down. I started to talk through it, "Well, it likely wouldn't change the downforce much because our downforce is mostly a function of the momentum change of the air." *must be the drag* "But it would increase the drag."

I stood still, trying not to have the, *is that right?* expression on my face.

Giving no indication of whether or not that was the right answer, he began pushing on the rear wing. "You have a lot of flex here, is that something you designed in?"

I gave a small laugh and said, "Nooo, no... We did point a camera at the wing to make sure there was not a noticeable amount of flex under normal loads. We didn't have the time to..." I saw his eyebrows rise and quickly decided to rephrase, "...or rather we chose to focus our time on other aspects of the aero system such as weight and packaging that would give a more overall benefit."

The argument that we didn't have the time to do something was never a good one. There was always some team that had some asshole with too much time who decided to do the most obscure and esoteric design exercise and all the judges want to know why we didn't do it too. The best answer, and the only real answer that isn't just an excuse, is that we chose to spend our limited time on something else that was a better use of our resources. It's true too, resource allocation is hugely important. Formula SAE is not an engineering competition; it is a program management competition. It just happens to require a lot of engineering.

Our discussion went on for a while until we were interrupted by the head design judge Mike O'Neil.

"That's all the time we have, I want to thank everybody for your participation and I need all the judges..."

I found myself standing next to Chris, and without looking at him, I asked, "So how'd it go?"

"We'll see," he said, with his serious face on.

"Now what?" Wes asked.

"Now... we go to the bar and get hammered drunk," I replied, still staring into the distance.

"Maybe we should wait until after the awards," Wes said.

"A little stressed, Matt?" Bobby asked.

"I don't know," I shook off my dazed stare, "my brain is just fried from thinking too much. I just need to, you know, not think."

"Does that mean you did well?" Wes asked.

"I don't know, he didn't really give me much feedback," I looked down at the gaping hole in the engine where a connecting rod decided to make an unscheduled exit. It was a stunning reminder that, in spite of the battle outcome, we had already lost the war. "I came to win and we didn't win, so, whatever. We made it to design finals, and that's not bad."

"Alright, let's take it back to the trailer and pack it up," Bobby said to the team.

I looked around at the parts I needed to take care of, and it dawned on me that it was all over. I had spent so much time the past week personally carrying the posters and aero parts, making sure everything was in perfect shape, and now there was nothing to take care of anymore. The wings were broken and the design posters were no longer needed; the car was done and our year was over.

"Who is the technical leader of your team?" I looked up to see Mike O'Neil talking to Brett. He pointed at me.

"This is what is going to happen," O'Neil continued, "I am going to ask all four of the technical leaders to come on stage and we're going to list off the order of the design finalists and ask you some questions about your car."

I nodded. "What kind of questions?"

"Just some questions about why you chose a design or what kind of testing you did. We also want the cars out in front so everyone can see them."

I nodded again.

"Alright? Good, we'll see you there."

I noticed Wes and Bobby had already left with the car. I chased after them and told them that we needed to bring the car over to the awards stage. The rest of the team went back to the trailer to start packing. Walking back, I noticed the mountains in the background. When we first got there, I was thinking about how strange it was to see mountains in the city. Once we got into the competition, it seemed like they had disappeared. Actually, they did a few times, when the smog from the city rolled in. But even when they were visible, we were all so focused on the car and the race that we lost the perspective to see mountains in the city even when they were so seemingly out of place. Auto racing does that.

"How do you think you did?" a voice came from behind me.

I turned around to see Lee carrying our fire extinguisher.

"Ummm... it's hard to say. All the other teams were so well prepared. I mean, Michigan State had a huge display board, BYU had a projector and a video on display."

"Kyle was saying that he thought the judges were spending most of their time with you guys."

"Yea, well the aerodynamics guy was for sure, but we were the only team with any wings. We'll see; I don't want to get my hopes up because I think it will be hard to overcome our engine failure. I mean they'll probably see that as a result of a design flaw."

"You guys are top four in design," he said, "and that's a pretty big accomplishment."

We went back to the trailer and packed up so we could leave after the awards. We were all in a weird mood. Anticipating the announcement of the design event results, but too tired to speak much more than a few words.

We went back to the awards stage and hung out around our car waiting for the design finalists review to start. We answered some questions and shrugged off the occasional "I'm sorry you guys didn't finish. That was a pretty spectacular failure though!"

"Okay, we're going to get started," Mike O'Neil announced over the speakers. "Can we get all the team technical leaders up here?"

I made my way on stage and stood on the far end next to Auburn University's Chief Engineer. I expected us to get third, though that may have just been wishful thinking.

"Back when I started FSAE, when most of you were still in elementary school, the level of competition was nowhere near what it is now. The teams today are at a whole new level, and I am amazed every year at what these teams can accomplish."

He talked about some different designs that teams were doing that the judges liked, and some that they didn't like. I stood nervously, wishing I had brought a bottle of water on stage with me.

"Our fourth place team surprised everyone. I don't think anyone expected them to do as well as they did..."

They certainly didn't. No one even noticed us until halfway through that event. Before that we were just another mid-level team.

He continued, "Brigham Young University was a first year team with a car that..."

Huh... I guess we did get third...

He talked about BYU's car, which I also thought was an amazing piece of engineering from a new team. He asked their guy a couple questions about their car and then continued, "Our third place team had some bad luck this year, dropping out of endurance with a car that was capable of winning the competition."

Well, I thought it was pretty good too; thank you for your kind words Mr. O'Neil...

He continued, "Michigan State University..."

Ha! We beat Michigan State? Michigan State! O'Neil asked them a few questions about their design and their engine failure in endurance. I wasn't paying attention, because I was busy trying to figure out how we had just convinced the judges we were better than Michigan State.

"And in second place, a great car yet again from the Auburn team. Again, your car is a solidly engineered car with plenty of testing, tell me about…"

I stood quietly with a huge smile trying not to distract from Auburn's moment. I wondered if it would be kosher for me to run around with a bottle of champagne spraying everyone and everything. Probably not.

After talking about Auburn's car for a few minutes, O'Neil walked over and stood next to me.

"All of the judges were impressed with the Oklahoma team this year. One of the outstanding features of the team was their cohesiveness. We all saw them working together like friends and getting along very well. They understood the car, they understood how each system interacts with and affects the other systems, and they had plenty of testing data and support to back up their design. Without a doubt, they had the best car. We felt like they were the best team even with their failure, which we expect was an occasional consequence of running a car near its limit and not an engineering or design failure. I think if we ran this competition five times, Oklahoma would win four of them. They showed the effectiveness of aerodynamics to a lot of apprehensive teams and judges. I expect there will be more teams with aerodynamics next year." He turned to me and asked, "Would you like to tell us about that failure you had on track?"

I stood still, nervously trying to think of something to say. After delaying for a few seconds, I said the only thing that came to my mind.

"Uhh… no."

Everyone laughed.

We ended up 14th overall. Impressive, considering that we scored zero points in endurance, which is worth 40% of the total event points. Texas A&M finished first place for the second year in a row. During the awards ceremony, we went on stage just about every other award. First place communications award, second place autocross event, and first place design event, among others. As we were leaving the awards with our cache of plaques and trophies, we heard a group of people behind us yell, "Boomer!"

Without even looking to see who it was, we all yelled, "Sooner!" and then turned around to see a group of students from Texas A&M.

It was good to know that we had at least made an impact, that we wouldn't be the only ones who remembered that great team who was at our trailer earlier that day, moments away from winning it all. Like a Cinderella team, we had come from nowhere to be in the spotlight on the dance floor. When midnight came, we didn't get our fairytale ending. But at least we got to dance.

Epilogue

A few of us stayed in California after the competition, as a little vacation. We drove around to different cities, saw the sights, and tacked on a few more memories to the end of our FSAE saga. In those few days, there was a lot of "Did we really win design?" and "I can't believe we actually got first place in design." It wasn't quite a consolation for not winning the competition, but it was pretty unbelievable to think that we had come that far. Adding up the scores, we found out that if we had finished endurance at the pace we started at, we would have won by almost 75 points. Pretty amazing considering most of these competitions have the third place team within single digits of first. We almost won, but how many teams could say that? Michigan State would have won if not for their engine failure just after ours. You always hear stories about people and teams that give everything they've got and end up winning or being the best. But for every team or person that wins against all odds, there are a handful of others with a story just as compelling, but who didn't have that last bit of luck they needed.

The design judges had all talked about our team's cohesiveness, how we all got along so well. At first I was amazed that we had been able to fool the judges into thinking we all got along at all. The team seemed to be almost split in two groups for half of the year, and we had more than our fair share of fights. But I started to think back on the year and how we all acted towards each other. We argued a lot, hated each other half the time, complained and said some mean things. But we also kept coming back, showing up to the next meeting, helping each other with homework, getting together at the bar those few free hours we had. We were like brothers. Like some fertile family had squeezed out a handful of engineers over the course of five years. In spite of our arguing and fighting, and in some ways because of it, Chris and I had been like brothers for almost as long as I had known him. Honestly, if you added up all the hours I spent with him over those four years it would probably be about the same amount of time I spent with my real brother over the previous 20 years. When you work with a group of people as closely as we had, for dozens and sometimes literally a hundred hours a week in continually stressful situations, you are going to want to hit them with a shovel at least a few times. But you also get to know them like they are family. That's what the judges saw, even when we couldn't see it ourselves.

When we finally got back to the shop, we unloaded our car and our trophies and went our separate ways. Kyle spent the summer working at a local go-kart track before going to England for a graduate degree in racecar engineering. We occasionally meet up when there is a racing event near one of us that the other has a good excuse to go to. Ricky had another year in school, but somehow managed to resist the pull of the race team to focus on graduating. Neither of us has a castle with a moat yet. Chris got married and went on to work for an automotive company in the same department as three other OU racing team alumni. After a few years, our paths crossed again, and we now work for the same company. Bobby remained on the team and was re-elected to team captain for a second term. To everyone's pleasant surprise, he eventually graduated. Wes graduated the same year as Bobby and went to work for an oil company in his home state of Texas. Superdave more or less disappeared only to be seen a few times around campus from a distance and in bad lighting, like a coffee addicted Sasquatch. He had the talent and focus to accomplish anything. I will always wonder what he ended up doing.

The next year, the team built a good car and finished all the events at both competitions, placing 21st in Detroit and 3rd in California. Every year, the graduating seniors would say, "This team is going to go downhill after we're gone" and every year we did better than the last. The 2008 team continued that trend; third place was the best finish the school ever had. Without being a prestigious school with heavy faculty involvement, it will take a lot of effort and a lot of luck for them to be a first place team. It will also take the right group of people together at the right time. I will always be looking for the Oklahoma team, hoping that at some time in the future there will be another group of kids there with the relentless passion and drive to win.

As for me, I got my interview with Claude and finally got my exciting career in auto racing. It had the racing and the engineering challenges, but it was missing some of the passion that drove me in FSAE. Professional racing is a different world, where the sponsors and the driver personalities are king, and the car is just along for the ride. FSAE is all about the car. It's like racing was fifty years ago, before the money and television coverage. That's all changed now in the business of professional racing. There are only a few places left where the money is thin and a trophy is the only prize. Those are the best kinds of racing, and FSAE is the best of the best.

All things considered, I think it's good to steer clear of the obsession that comes with an unfettered drive to win. Racing is addicting, it has a model of intermittent reinforcement like gambling; you usually lose, but occasionally you win just enough to keep coming back. It will take over your life if you let it, pushing everything and everyone else away. As a node, a part of your larger life, it's hard for me to imagine something so exciting and interesting. Within the passion of fierce competition, within life's school of war, there are friends and family and experiences that make it all worth it, even if the luck and the trophies never come.

Me in the computer lab, updating the schedule. Stars and cars to be added later.

Bobby running the CNC mill in our bomb shelter/machine shop.

Above: Our shop with the 2006 car in the foreground, '05 in the background, and previous year car bodies hanging on the wall. *Below:* Frame manufacturing begins, and Wes is eager to test out the ergonomics.

Ricky and Andy delicately make the necessary modifications to the engine case.

The car begins to look like a car. Kind of.

Andy testing the
torsional rigidity of
the frame.

Right: It is always
important to keep a
clean work area.

Below: Ricky and
Chris bench testing
the electronics
before installing
them on the car.

11:55 PM Mountain Standard Time.

Chris and Bobby doing cooling system testing in our basketball stadium parking lot on a quiet weekend.

Bobby doing some testing with our cutting edge impact attenuator, AKA road cone.

Kyle and Chris doing some wingless testing at the commercial building complex around our shop.

Right: No Sleep Till... Pretty much right after we leave!

Below: That's me getting in the zone just before my first autocross run in Detroit. The cowboy hat is standard issue to anyone living in Oklahoma.

Lee is so happy that he no longer has to drive around the greater Detroit area searching for exhaust parts.

Pre endurance. Notice the sweet pushbar. Also notice how it's all about Chuck.

California mountains starting to disappear in the distance.

Sprocket hub. You can see cracks between the small holes and the larger half circles where the sprocket attaches.

Why yes, that is a pirate ship themed kids pool in our paddock.

8 University of Oklahoma

Team Name: Sooner Racing Team

Faculty Advisors: Mitch Burrus, Zahed Siddique

Sponsors: Universtiy of Oklahoma, 3M, Schlumberger, Napa, Iron Horse Developement, Stone Canyon Exploration, Compressco, Burrus Family

BODYWORK: Rich Mahogany
BRAKES: The Midas Touch
COLORS: Rich Mahogany & Bird
COOLING SYSTEM: Water
DRIVE: Hope and Inspiration
ENGINE: Yes
FRAME: Bamboo strapped w/Vines
FRONT SUSPENSION: Single Axle
FRONT TRACK: Rich Mahogany

FUEL SYSTEM: TMI (Tuned Monkey Injection)
FUEL TYPE: Sunoco
INDUCTION: Air
OVERALL HEIGHT: 5
OVERALL LENGTH: ask your mom
OVERALL WIDTH: 34"
REAR SUSPENSION: Posi
REAR TRACK: Rich Mahogany
SHIFTER: Minty Fresh

TIRES: Round
UNIQUE FEATURES: Many Leather bound books, Cup Holders, Innovative Innovations, and of course- Rich Mahogany
WEIGHT: We weight for no one
WHEELBASE: At the "Wheel-Lair"

All the important, need to know stats of the 2007 Oklahoma FSAE car.

Far Left: The team and car during design semi-finals.

Near Left: Kyle and I, along with one of our Kansas friends, Nick, walking the endurance course before the event.

Bottom Left: Kyle and I discussing the finer points of driving fast.

Top: All the teams gathered together for the panoramic photo.

Below: Superfast Matt, doing my thing during the endurance race.

Kyle making the tires work. Oklahoma gives you
lateral acceleration.

Our smoldering wreckage.

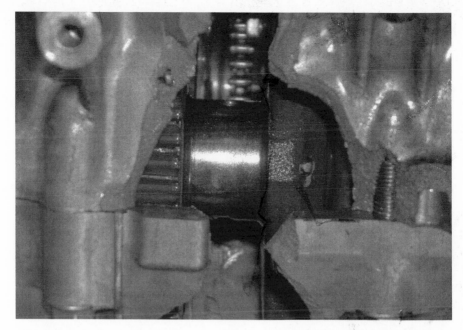

The big hole in the bottom of the engine where the connecting rod embarked on its solo career.

From left to right, back row: Lauren, Wes, Ricky, Josh, me, Chris, Kyle, Phil, Dr. Siddique, Mitch, and Dr. Gollahalli. Front row: Brett, Bobby, and Trey.